The Seaweed Handbook

An Illustrated Guide to Seaweeds from North Carolina to the Arctic

by Thomas F. Lee

Illustrations by Wendy Webster

Dover Publications, Inc., New York

Published in Canada by General Publishing Company, Ltd., 30 Lesmill Road, Don Mills, Toronto, Ontario.
Published in the United Kingdom by Constable and Company, Ltd., 10 Orange Street, London WC2H 7EG.

This Dover edition, first published in 1986, is an unabridged republication, corrected by the author, of the work originally published by The Mariners Press, Boston, Massachusetts, in 1977.

Manufactured in the United States of America
Dover Publications, Inc., 31 East 2nd Street, Mineola, N.Y. 11501

Library of Congress Cataloging-in-Publication Data

Lee, Thomas F.
The seaweed handbook.

"Unabridged republication, corrected by the author, of the work originally published by The Mariners Press, Boston, Massachusetts, in 1977"—T.p. verso.
Bibliography: p.
Includes index.
1. Marine algae—Atlantic Coast (U.S.)—Identification. 2. Marine algae—Atlantic Coast (Canada)—Identification. 3. Marine algae—Arctic Coast (Canada)—Identification. I. Title.
QK571.L43 1986 589.4'5'0974 86-16562
ISBN 0-486-25215-9 (pbk.)

"There is a pleasure in the pathless woods,
There is a rapture on the lonely shore,
There is society where none intrudes,
By the deep Sea, and music in its roar:
I love not Man the less, but Nature more,
From these our interviews, in which I steal
From all I may be or have been before,
To mingle with the Universe, and feel
What I can ne'er express, yet cannot all conceal."

Lord Byron

"I have heard the mermaids singing, each to each.

I do not think that they will sing to me.

I have seen them riding seaward on the waves
Combing the white hair of the waves blown back
When the wind blows the water white and black.

We have lingered in the chambers of the sea
By sea-girls wreathed with seaweed red and brown
Till human voices wake us, and we drown."

T.S. Eliot

"Men who fish, botanize, work with the turning-lathe, or gather seaweeds, will make admirable husbands."

Robert Louis Stevenson

to M.B.L. '67

Contents

Introduction

It is paradoxical that while there is so much water around us, we still know so little about what lies in its depths. The sea is our last frontier, and many of its inhabitants are still waiting to be discovered and appreciated. This book has been written to further that exploration by introducing the reader to a world of beautiful, abundant and important living organisms—the seaweeds. It should be particularly useful to students, amateur naturalists, beachcombers and divers, or anyone who is not aware of the richness and diversity of plant life beneath the coastal waters.

Hundreds of individuals complete S.C.U.B.A. courses annually, but many give up diving within a few years. They become bored with such obvious activities as spear-fishing and wreck-diving but lack the knowledge to utilize their hobby in other ways. They do not realize that diving, particularly in the oceans, enables them to view a world of plants which most human beings never see or even imagine. Along the U.S. East Coast the variety of sea plants is tremendously diverse, but like the hiker passing through the forest unaware of the identity, or even sometimes the presence of the trees and flowers around him, the average diver glides over forests of seaweeds with scarcely any notice.

The seaweeds are readily available to others as well. A walk along a rocky coast at low tide will reveal to the interested observer dozens of species of these plants, either growing on rocks and shells or simply being carried shoreward by the moving water, having been detached from the depths. Vigorous wave action tosses up onto the beach great mounds of seaweeds, in which careful searching will reveal numerous species.

HOW TO USE THIS BOOK

These preliminary pages supply the background necessary for understanding what the seaweeds look like and why, where and how one can find them, their vital role in the life processes in the oceans, how to collect, keep and even (sometimes) eat them, and other facts about these remarkable plants that will enable the reader to approach their observation and study within a useful frame of reference.

In the second section of the book, 78 species have been selected for individual treatment, and each of these is illustrated as it appears in nature, with an accompanying description which includes pertinent and useful information such as the scientific name and common name (if any), and the size, color and location. The plants have been arranged according to structure. In other words, rather than group them according to their precise scientific classification (p. 13) those seaweeds which have the same basic body plan such as the membrane, blade, etc. are treated together. Simplified "keys" for identifying the specimens are found at the beginning of each section. A key consists of a series of choices which, when followed carefully, will lead one to the proper illustration and description. Even though the structure of these organisms may be somewhat variable, which will perhaps necessitate leafing through several illustrations in order to find the plant in question, this grouping is actually rather consistent, and the practice of examining additional illustrations as a regular exercise will soon result in an increasing familiarity with a wide variety of seaweeds.

WHAT GEOGRAPHICAL AREA DOES THIS BOOK COVER?

Seaweeds, like the land plants, are affected by their surrounding ecological conditions. The geographical distribution of specific seaweeds is governed by such factors as water temperature, salinity and light intensity. As a result, there are several distinct although overlapping populations of these organisms on the shores of the United States and Canada.

This book concerns itself with that large group of seaweeds found from Beaufort, North Carolina northward, and extending along the

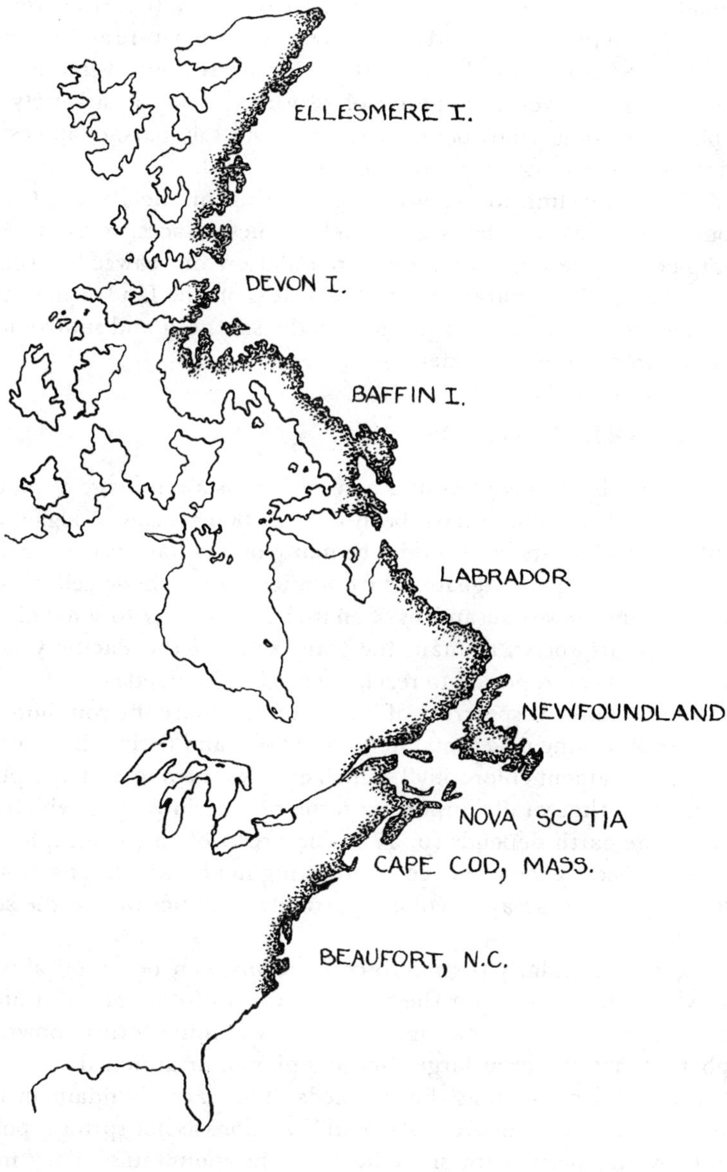

Figure 1. The shaded areas represent the extent of the geographical area covered in this book.

Canadian Maritime Provinces all the way to the Arctic (fig. 1). Seaweeds southward of this region tend to be more tropical in nature and are thus outside the scope of this book. In the extensive region covered in this book there are over 400 species of seaweeds, living in a variety of habitats including sandy bottoms, rocky coasts, salt marshes and estuaries, and extending far out on the continental shelf.

A diver traveling to the waters of Florida or the Pacific Coast would find himself or herself, in each of these places, in an almost completely different and unique population of seaweed species. This diversity is as marked in the sea as it is on the land, where the northern temperate forests give way to the southern and subtropical vegetation of lower latitudes.

WHAT ARE SEAWEEDS?

The seaweeds are members of a heterogeneous assemblage of plants called the algae. There have been over 25,000 species of algae described by scientists, and no doubt many others have yet to be discovered. The algae range in size from microscopic single cells which may be only a few thousandths of an inch in diameter to some of the largest plants known to man, the giant kelps of the Pacific Coast, which have been reported to reach a length of 150 feet!

Despite this wide spectrum of form, they all share the common attributes of lacking true roots, stems or leaves, and their cells contain the green pigment chlorophyll, which enables them to undergo photosynthesis, that vitally important series of reactions upon which all life on the earth depends (p. 17). The green of this chlorophyll is often masked by the presence of other pigments, which lend to the algae a dazzling array of colors, particularly noticeable in the seaweeds.

Algae, particularly the microscopic forms, can be found almost anywhere one searches for them. They grow in fresh and salt water, either as microscopic floating or weakly swimming forms known as "phytoplankton," or as larger floating plants, or attached as larger "benthic" forms, such as the seaweeds. They are abundant in the soil, and may occur in such "stressful" locations as hot springs, polar ice caps and permanent snow fields in the mountains. They may cover rocks and trees, usually as a slimy green coating, and some live as epiphytes, that is, attached to the surfaces of other plants or animals. Others are endophytes, living within the tissues of other organisms such as protozoans, sea anemones and corals. The algae can also be "airborne." A recent research project on the summit of a New Hampshire mountain resulted in the isolation of 23 species of microscopic algae from the air!

It is estimated that algae were present on the earth well over two billion years ago, and probably all other plants have since evolved from those ancient ancestors.

The seaweeds are specific kinds of algae, and are macroscopic (visible without the aid of a microscope) plants which grow attached in the oceans. They are often referred to as "marine algae," a term which also includes the phytoplankton. They have long fascinated the amateur naturalist as well as the professional scientist, and with the invention of S.C.U.B.A. the opportunity has been presented to observe and study these plants in their natural surroundings.

WHAT DO THE SEAWEEDS LOOK LIKE, AND WHY?

The form of the seaweeds, like that of any plant or animal, is the result of millions of years of evolutionary development, shaped by the forces of the environment. The simplest form is that of a *single, spherical cell.* Down in the tropics, some of these single celled plants may grow as large as 1 inch in diameter.

Many seaweeds assume the form of very thin threads, or *uniseriate filaments.* These are essentially chains of cells, attached end to end like a string of beads. The individual cells may be too small to see without magnification, such as those in *Calothrix* (plate 71), or they may be quite conspicuous, as in *Chaetomorpha melagonium* (plate 36). The uniseriate filament may also be branched, like *Cladophora* (plate 58) or *Ectocarpus* (plate 63).

When the filamentous character is very complex and consists of multiple rows of cells and branches, the seaweed becomes larger and may form bushy tufts or clumps, such as *Polysiphonia* (fig. 2) or *Ceramium* (plate 66).

Others have developed as large sheets of cells, or *membranes,* which ordinarily are 1 or 2 cells in thickness. Common examples are the bright green "sea lettuce" *Ulva* (plate 3) and the edible *Porphyra* (fig. 3).

Quite often, the diver may see large areas of rocks and shells cov-

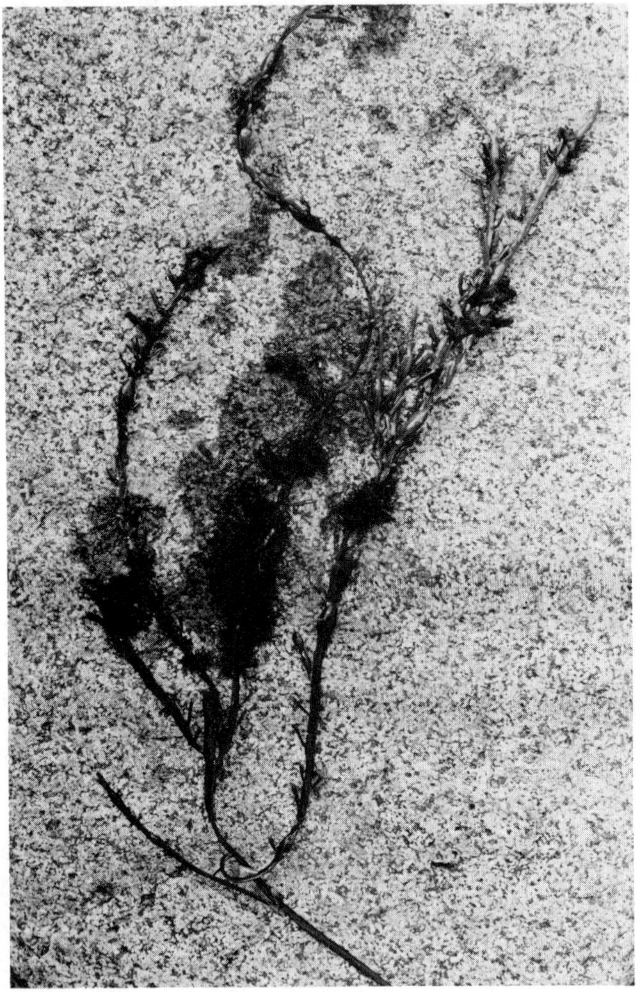

Figure 2. *Polysiphonia lanosa* forming bushy clumps on *Ascophyllum nodosum*.

Figure 3. ***Porphyra umbilicalis*, an edible red algal membrane.**

ered with pink or red or other colorful "crusts" (plates 72-74). These consist of layers of microscopic filaments clinging to the substrate and supporting erect rows of colored filaments, often encrusted with a hard outer layer of calcium and magnesium carbonate.

The largest and most complex of the seaweeds are those which may, for the purposes of this book, be called the *tissue* type. These are made up of millions of cells, arranged to produce large and often very strong plants such as *Fucus* (fig. 4) or *Laminaria* (plate 11). Others form tubes, such as *Scytosiphon* (fig. 5), or cords, such as *Chorda* (fig. 6). These are the seaweeds that one first notices covering the rocks and jetties along our coastline, or finds tossed up on

Figure 4. *Fucus distichus*, a branching, "tissue" type seaweed.

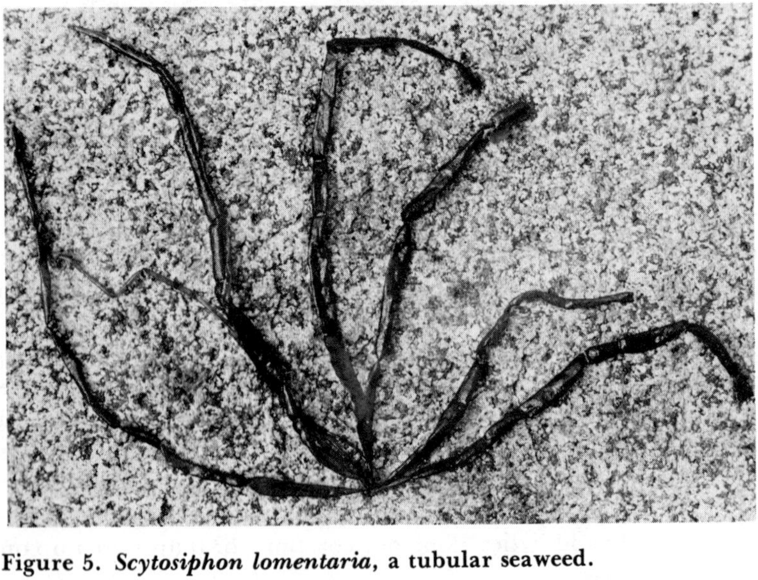

Figure 5. *Scytosiphon lomentaria*, a tubular seaweed.

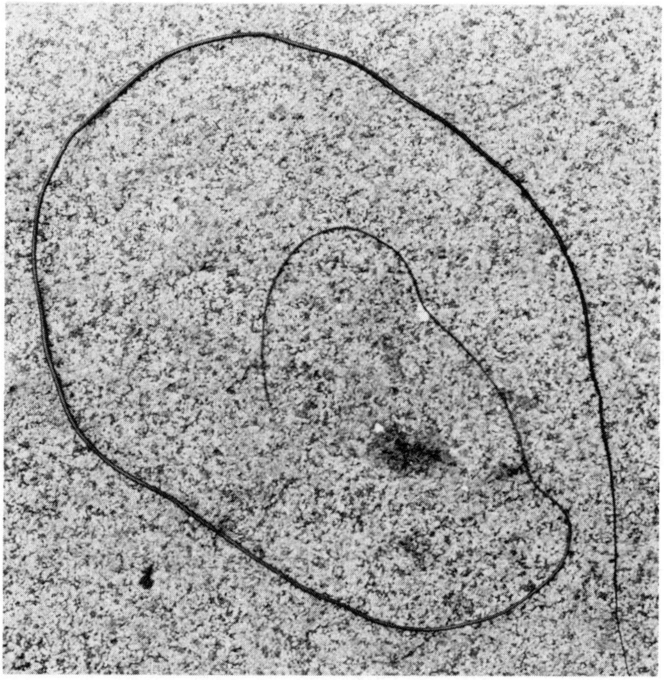

Figure 6. ***Chorda filum,*** **a cordlike seaweed.**

the beach after a storm. Many of this type grow completely submerged, and their grace and beauty as they move back and forth in the ceaseless underwater ebb and flow of the sea can only be fully appreciated by the diver who visits their natural habitat.

The shape assumed by the seaweeds is an expression of their biological needs. Plants growing in water do not need the ordinary roots, stems and leaves of land plants. The latter need roots for attachment and absorption of water and mineral nutrients from the soil. The seaweeds, immersed in their own water and nutrient supply, absorb these materials directly into their cells, and need only secure attachment by a primitive rootlike device, the "holdfast." Figure 7 demonstrates the body plan of a representative seaweed. The holdfast may be quite small and consist of a tiny disc, or a few filaments, or may be quite large and elaborate, depending on the stresses to which the particular plant is put.

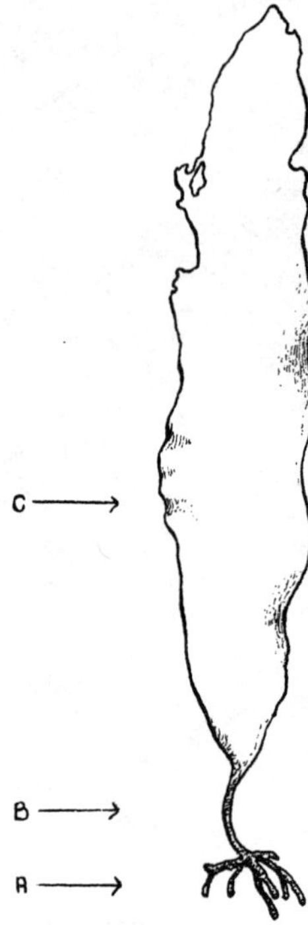

Figure 7. ***Laminaria*, a common seaweed. (A) the holdfast, for attachment (B) the cylindrical stipe (C) the flat photosynthetic blade.**

The land plant needs a firm stem to hold itself upright and transfer the material absorbed through the roots to the leaves. The seaweed is buoyed up by the water, and has no need of elaborate conducting tissue although some larger genera have a primitive conduction system. It would be a disadvantage for a seaweed to have a stiff stem, since the sometimes violent movement of the sea would soon break such a structure. Certain species do have a stipe, a cylindrical flexible shaft rising from the holdfast and supporting the remainder of the plant (fig. 8). Also, air bladders in some assist in the flotation of the photosynthetic tissues (plates 22, 76).

Photosynthesis on land is carried out in leaves. These are replaced in the seaweeds by a complex array of various types of blades, membranes

Figure 8. ***Laminaria saccharina.*** **Note the elaborate holdfast and flexible, cylindrical stipe.**

and filaments, most of which lack any conducting tissues, and carry on the reactions of photosynthesis very efficiently using materials extracted directly from the surrounding watery medium.

HOW DO THE SEAWEEDS REPRODUCE?

While a detailed description of the reproductive processes of the seaweeds is not vital to the purposes of this book, the identification of many species is based upon the recognition of their reproductive structures. It is also important to know something of the life history of the organisms, since their habits of propagation may cause them to be found only during certain months of the year.

For example, the several species of *Fucus,* the common brown "rockweed," are easily separated by the particular shape of the large, swollen, granular structures called "receptacles," located at the tips of the plant (fig. 9). These receptacles contain thousands of sperm and egg cells which, after their release from the plant at low tide may unite to form new plants. Survival of these reproductive cells and the young plants in the sea is of course subject to many hazards

Figure 9. Receptacles of *Fucus spiralis*. They contain the sperm and eggs for reproduction.

such as grazing by animal predators, and very few reach maturity. Those consumed are not to be considered wasted, but make a valuable contribution to the sustenance of animal life in the area.

Reproductive cells which do not fuse like sperm and egg cells but can simply grow by cell division into a new plant or alternate phase in the life cycle are termed zoospores. Specialized structures have evolved in some seaweeds to produce these zoospores, such as the small blades growing from the stipe of *Alaria* (plate 5). The large brown seaweed *Laminaria* (plates 7, 11-13) produces these zoospores all along the edge of its blade, and the membrane of *Ulva* (plate 3) in its reproductive phase has a distinct bleached appearance along its edges due to zoospore release.

To make matters even more complicated, some of the seaweeds have two or three phases in their life history, and each phase may look completely unlike the others. For example, the large membrane of *Porphyra* (plate 1) may exist for part of the year as a microscopic red filament growing on shells, whereas the common branched *Polysiphonia* (plate 62) has 3 distinct phases, all of which look alike but which are male, female or a combination of both! Of course, algae never produce seeds, which have only evolved "recently" in the conifers and flowering plants.

Reproductive characteristics are not emphasized in this book, but are referred to where necessary. The features by which most seaweeds are identified are readily visible with the unaided eye, although occasionally it will be useful to use a hand lens to see some of the smaller characteristics.

HOW ARE SEAWEEDS NAMED?

Living organisms are placed into categories in order to indicate their evolutionary relationships with one another. That is, all plants in one particular category share common ancestors, much as all of the people in a particular family, clan or tribe are descended from the same individuals.

The science of classification of living organisms is called Taxonomy, and the largest and most comprehensive taxonomic group within the plants is called the Division. This is analogous to the Phylum of the animal world.

Thus, there are some seven Divisions of algae, four of which number seaweeds among their members. A Division name always ends in the suffix phyta, from the Greek word meaning plant. Divisions which include the larger, visible seaweeds are the Cyanophyta (blue-green algae), Chlorophyta (green algae), Rhodophyta (red algae), and the Phaeophyta (brown algae). It should be noted that the Cyanophyta are now technically considered to be specialized forms of bacteria, although they are often still referred to as algae.

Each particular plant also is given two names in order to distinguish it from all others. These are the genus and the species. They are usually derived from Greek or Latin words, and therefore can be used by scientists from all over the world without translation into any other language. To illustrate this using a well-known land plant, the Sugar Maple is known scientifically as *Acer saccharum*. The genus *Acer* means maple, and is always capitalized. The species *saccharum* means sugary and is never capitalized. One genus may have several species, so there is an *Acer platanoides* (Norway Maple) and an *Acer nigrum* (Black Maple), etc. When a plant is cited by its genus name only, without indicating a particular species, the genus name may be followed by sp., an abbreviation for species. For examples in this book see pages 162 and 168. It should be pointed out that there are classification categories at levels lower than the species, such as subspecies, but we will not refer to them in this book.

Many of the seaweeds have common names, such as "sea lettuce" (*Ulva lactuca*) or "laver" (*Porphyra umbilicalis*), but these differ from place to place so are not used for scientific purposes.

Although the names of the seaweeds may initially seem complicated and difficult to remember, after the repetition of learning to associate these names with specific plants they will become surprisingly familiar.

Listed below are those Divisions of algae which include the seaweeds treated in this book along with a brief description of the major characteristics of each Division.

1. Division Cyanophyta (blue-green algae).

These occur in salt water mainly as conspicuous, slimy blackish masses of microscopic filaments on rocks just at the high tide level, and on the surface of mud flats in salt marshes and estuaries. The color of the Red Sea is due to massive growths of the floating genus *Trichodesmium*.

2. Division Chlorophyta (green algae).

These will be seen as branched and unbranched filaments, and membranes. They are usually a very obvious bright green, and are not commonly found below a depth of 40 feet. When brown or red algae are found detached or thrown up on the shore they may be mistaken for green algae because their darker pigments have disappeared due to the degeneration of the plant. Some of the green algae such as *Enteromorpha* (plates 2, 14) have adapted to growing in areas of low salinity, so they are conspicuous near river mouths and other fresh water discharges.

3. Division Phaeophyta (brown algae).

These are the most conspicuous and abundant of our seaweeds. They are generally large, "tissue" type plants and cover the rocks, jetties and bottoms of our colder, shallow offshore waters. Although the East Coast species do not attain the length of the giant Pacific kelp, which has been reported to reach 150 feet, they often grow as long as 10 or 20 feet and form dense beds, such as *Laminaria* (fig. 10).

Figure 10. ***Laminaria longicruris*, a very large kelp. This specimen is from northern Maine waters.**

4. Division Rhodophyta (red algae).

These are of a smaller and more delicate construction than the brown algae, but are abundant and readily seen. They range in color from bright red to purple to some that appear quite brown. They are among the most strikingly beautiful of all the seaweeds (fig. 11). The red algae are common at all depths, and become the dominant plants in the deeper waters down to 100 feet or more along our coast, and in the tropics to about 875 feet. The hard, crustlike plants which cover rocks and shells are members of this Division. Some of the red algae are edible and commercially valuable.

Figure 11. A beautifully delicate red seaweed, *Phycodrys rubens*.

ARE THE SEAWEEDS IMPORTANT?

The ecology of the oceans is complex, but one thing is certain. The continuation of life in the oceans depends upon the presence and growth of the sea plants. These plants are the algae, and include, as mentioned earlier, both microscopic forms and the larger, attached seaweeds. These algae are of such vital importance because they are capable of carrying out the process of photosynthesis. Living organisms need energy, and the ultimate source of energy is the sun. Photosynthesis is the only series of reactions which can convert the light energy of the sun into chemical energy, and plants alone are capable of doing this. Carbon dioxide and water are combined during this energy conversion in the cells of the plant to become organic matter such as sugars or fats, and oxygen is released as an important by-product. It is estimated that the sea plants may produce a significant percentage of the total oxygen released by all plants on the earth each year.

This means that although the plants themselves use up some of the organics and oxygen for their own life processes, the rest is available for use by all of the other living creatures in the oceans, from the smallest microscopic invertebrates to the whales. If the algae were to disappear from the oceans, so would most of the living creatures. And since the rate of photosynthesis is dependent upon a number of factors in the environment, these factors aid in determining where the plants and consequently the animals will be found.

A good example is the formation of coral reefs. These are built up both by the action of millions of tiny animals, the coral polyps, who secrete about themselves a skeleton of calcium carbonate, and by the growth of red seaweeds, whose forms are covered with this same material. Many of the coral polyps have microscopic algae growing within their tissues, and these tiny plant cells pass energy-giving food directly to their polyp hosts. Since photosynthesis is dependent upon sunlight, these reefs will not form at depths where the sun's rays do not penetrate, so the formation and distribution of these coral reefs is ultimately dependent upon the photosynthetic activity of the algae.

On a larger scale, the average depth of the oceans is almost 2 miles. Sunlight, under ideal conditions, does not penetrate more than about 600 feet into the water. In fact, even in very clear water at only 60 feet, the light intensity has already decreased by half, and only 10% reaches depths of 150 feet. Cloudy coastal waters have much less light penetration. Off Cape Cod, for example, half the light is lost in the first 12 feet, and at a depth of about 55 feet, only 1% of the light remains.

This means that below this "photic zone," or zone of light penetration, constant darkness prevails. Most of the creatures living in this eternally dark, cold and relatively unexplored environment must exist on food materials which come down to them from the upper, richer waters. There is a continual slow raining down of the phytoplankton to these depths. Certain members of the phytoplankton, the one-celled diatoms, have a hard, glasslike covering of silica, and in some areas the descent of the remains of these plants has, over the centuries, built up sediments hundreds of feet deep. There are sites, such as California, where the sea has receded, and these sediments are mined and sold as "diatomaceous earth," valuable as an abrasive and filtering agent.

The attached seaweeds can enter into this energy transferring food chain in the following manner. As they mature and die off, they are partially digested by marine bacteria. These in turn are eaten by small animals, who further digest the seaweed material. What they cannot use is passed out, and becomes reinfested again by bacteria, and so the process continues. These small animals are food for fish and other larger animals. There are of course numerous animals such as sea urchins and periwinkles which consume the seaweeds directly as a major part of their diet.

To emphasize the importance of the seaweeds, recent calculations have estimated that perhaps at least one half of the total weight of all living matter on the earth is made up of marine algae! Also, it has been discovered that these algae are the most productive of any of the plants. For example, beds of *Laminaria* (plates 7, 11-13) along the Canadian Atlantic Coast produce more organic material per year than a tropical rain forest or even a highly cultivated field of alfalfa. *Fucus* (plates 18, 20, 21) and *Ascophyllum* (plate 22) can actually double their weight in five to ten days. At the edge of the sea then, the seaweeds are the sites of intense productivity, and help create conditions necessary for the abundant growth of other organisms, upon which young fish feed, and thus they help to form the basis of many of our commercial fisheries.

WHERE ARE SEAWEEDS FOUND AND WHY?

A wide variety of ecological factors determines where seaweeds as a group and also particular species may be located. None of these factors act alone but their combined effects result in a predictable and specific distribution of these plants. A basic understanding therefore of these environmental pressures acting on the seaweeds is necessary if one wants to appreciate the marked differences seen in the populations of seaweeds as one dives and explores various localities along the coast.

1. Light.

The large attached seaweeds live within the photic zone, and hence are confined, as a rule, to the continental shelves. While a species of a red alga has been collected from a depth of 875 feet in the Bahamas, most seaweeds in our particular area are limited to a depth of less than 150 feet. This is because the water is relatively cloudy from being stirred up by such forces as wind, currents, river mouths, etc.

It is not only the amount of light present, but also its "quality" or color, which regulates seaweed growth. The sun's rays are made up of different wavelengths of light, some of which, in the visible range, appear colored. Plants use those wavelengths from the shorter blue rays to the longer red wavelengths in the act of photosynthesis. Ocean water is much less transparent to red light than to blue, thus, most of the orange-red components are quickly absorbed in the upper levels of the water, while the blue light penetrates to the greatest depths. This partially explains why very clear, deep oceanic water appears quite blue. It also explains that general trend in seaweed distribution which results in the red algae (which absorb mostly blue light) being found at the greatest depths, while the browns are more intermediate in position, and the green algae (which absorb mostly red light) are found nearest to the surface. The preference for different wavelengths results from the fact that these seaweed Divisions have different pigments in their light-trapping cells. This adaptation of specific wavelengths helps explain why the seaweeds have such a bewildering array of colors.

This general trend of depth distribution is not to be applied to all species. Nature has exceptions to almost every rule, so that many red algae, for example, can be found intermingled with the green seaweeds growing in shallow water. Also, many of the shallow, rock-covering seaweeds are members of the brown algae. Even so, one will find that the general sequence of green-brown-red is often quite apparent.

2. Nutrients.

While limited in depth due to decreased light penetration, the seaweeds (and the phytoplankton) grow profusely over the continental shelves due to the presence of abundant essential plant nutrients, principally nitrogen and phosphorus. Just as addition of these to soil increases the productivity of a garden, an increase in their concentration in the ocean may have a fertilizing effect on the growth of the algae.

These nutrients slowly leach out of the soil, enter the streams and rivers, and eventually pass into the sea. It is theorized that periods of excessive rainfall may flush enough of these nutrients into the sea to cause, in conjunction with temperature and salinity changes, such a rapid population explosion of certain phytoplankton that the phenomenon of "red tide" may occur. This is a burst of growth of a population of millions of cells of the microscopic genus *Gonyaulax*, which appears as great patches of brownish-red waters. Filter-feeding shellfish such as clams or oysters concentrate these cells in their tissues, and consumption of this toxic tissue may paralyze or even kill unsuspecting humans.

The seaweeds too flourish in this nutrient-rich water. Since sewage carries with it large amounts of nitrogen and phosphorus, the paradoxical situation may develop where sewage may actually increase the growth of seaweeds! What usually occurs, however, is that the load of sewage becomes excessive, and literally covers the plants and the bottom, and all life but the most hardy is wiped out.

3. Temperature.

Temperature is a very important factor in the distribution of the seaweeds. Photosynthesis uses carbon dioxide, and colder waters are able to hold higher concentrations of dissolved carbon dioxide than the warm tropical waters. Also, some seaweeds have developed methods to adapt their metabolism to cold temperatures. Thus we have the somewhat surprising fact that seaweeds grow very rapidly and in great numbers in colder waters such as the North Atlantic, and do not flourish in the same way in the warmer waters of the tropics. Out on the West Coast, a very rich flora develops in the cold Pacific waters. Off Monterey Peninsula, California, more than 400 species of seaweeds have been reported. This is approximately as many species as we have on the entire northeastern North American Coast!

The observant diver may find a wide variety of seaweed species when diving in the Florida Keys, for example, but could find a more luxurious growth of fewer but much larger species off the coasts of Massachusetts and Maine. There is evidence as well that photosynthesis may even be more efficient at lower temperatures in the marine algae.

The annual variation in water temperature has a marked effect on the geography of particular species of seaweeds. In the tropics this variation may amount to only 2 or 3 degrees Centigrade, but along the Atlantic Coast of North America it may vary as much as 18 degrees Centigrade. Temperature also varies with depth, and on shallow bottoms much greater fluctuations may occur than in deeper waters.

Along our North Atlantic Coast the warm waters of the Gulf Stream moving northeastward from the Caribbean have a major influence on the air and water temperature over the continental shelf. Cape Cod represents a north-south boundary for many species affected by water temperature, since the waters north of the Cape are chilled by the Labrador Current and species thriving in the warmer waters south of the Cape cannot tolerate the lower temperatures of Massachusetts Bay northward.

For example, organisms such as *Sargassum filipendula* (plate 76), *Bryopsis plumosa* (plate 57) and *Codium fragile* (fig. 12) are seldom found north of the Cape, and *Laminaria longicruris* (plate 12), *Agarum cribrosum* (plate 4) and *Callophyllis cristata* (plate 28) are not found to the south of the Cape, except possibly in some colder water stations such as the eastern tip of Long Island.

Other species such as *Fucus vesiculosus* (plate 18) and *Stilophora rhizodes* (plate 42) extend from the Canadian Maritimes down to Beaufort, North Carolina, the southernmost limit of the geograph-

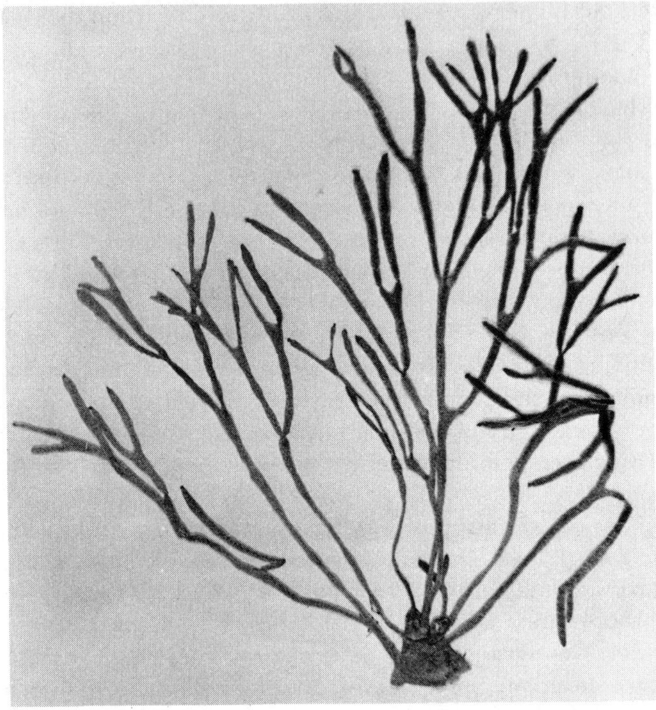

Figure 12. ***Codium fragile*, a green ropelike seaweed, whose growth is generally restricted to waters south of Cape Cod and north of New Jersey.**

ical area covered by this book. Others, like *Ulva lactuca* (plate 3) and *Ascophyllum nodosum* (plate 22) grow from the Arctic to the tropics.

From North Carolina to Florida tropical seaweed types become increasingly common, and a tropical flora dominates southern Florida, the Keys, and the Dry Tortugas.

As each seaweed species is described and discussed in the latter part of this book, the geographical distribution will be given, insofar as it is currently known. It is important to realize that the reported distribution of seaweed is often related to the amount of searching that has been done and may sometimes be changed as new areas are explored.

4. Salinity.

The salinity, or concentration of salts, in ocean water ranges from near 0% in areas of fresh water run-off to 4% or more in specialized environments such as the Red Sea. However, it is actually quite constant in most waters and is usually somewhere between 3.4 and 3.7%. A good example of a seaweed which can tolerate almost all levels of salinity is the common green genus *Enteromorpha* (plates 2, 14). Along river mouths or even where a fresh water stream trickles into the ocean, the bright green strands of *Enteromorpha* stand out. As a result of this relative indifference to salinity, *Enteromorpha* flourishes well up into the rivers, estuaries and salt marshes and may cover pilings and ship bottoms, sometimes causing severe fouling on the latter.

Most seaweeds, however, have narrow salinity tolerances, and would be killed by the addition of too much fresh water, or by the evaporation of water and subsequent concentration which occurs in areas such as tide pools. These pools represent specialized habitats, and can fluctuate in a day's time from being a dilute, cool deep pool to a concentrated, warm shallow environment. Here again, *Enteromorpha* will be found, as well as *Ulva* (plate 3), *Cladophora* (plate 58) and others.

5. Currents and Wave Action.

In wave-tossed waters such as those of the rocky headlands of Cape Ann, Massachusetts, Boars Head, New Hampshire, or in the swift, cold currents near Sandwich Beach, Cape Cod, or Passamaquoddy Bay, Maine, certain species of large brown algae flourish. For example, the edible *Alaria esculenta* (plate 5) is found where there is strong wave action, and rich growths of *Laminaria digitata* (plate 7) are found in currents strong enough to detach less resistant species.

In contrast, plants like *Ascophyllum nodosum* (plate 22) favor more sheltered shores. Apparently, plants like *Alaria* or *Laminaria* have the structural toughness and flexibility to withstand these environmental stresses, and are able to produce reproductive cells which can stick to a surface long enough to develop into resistant adult plants.

6. Substrate.

The observant diver moving over the sandy bottom off Plum Island or the New Jersey shore may wonder whether or not the descriptions he has read here of the rich seaweed populations of the North Atlantic are accurate. The bottom is bare except for occasional patches of detached plants drifting just above the sand. This points out another factor which governs the distribution of the seaweeds. A suitable substrate, or surface, is necessary so that the seaweed can attach itself and grow, fed by the water currents and exposed to the sun's rays. Just as a germinating seed on land needs the right soil in which to extend its roots and collect nourishment, the seaweed needs an appropriate surface on which to attach its holdfast (fig. 13).

As a general rule, surfaces which are somewhat rough and stable are utilized. These would include surfaces of rocks, old shells, wharf pilings, jetties, and any objects lying on the bottom such as lobster pots. Also, once the larger forms establish themselves, smaller species can use them as places of attachment, so that careful exploration around the bases of *Laminaria* (plates 7, 11-13) for example may reveal a variety of the smaller delicate red algae.

Figure 13. **Rocky coasts, like this one in southern Maine, afford good surfaces on which seaweeds can attach.**

ZONATION

The discussion so far concerning the distribution of seaweeds along our coast has concerned itself with the major ecological factors that influence the character of seaweed populations. Having outlined these major trends in seaweed geography, it will be very useful to point out more examples of where specific algal Divisions, genera and species described and illustrated in this book are apt to be encountered.

The language of ecology is sometimes rather confusing, and ecologists differ on the use of certain terms to describe particular habitats, or "homes," of plants and animals. As a result, the terms used here are chosen for their simplicity and convenience. They describe the important phenomenon of marine biology known as "vertical zonation." This means that species of organisms tend to group together in "bands" or "stripes" which run roughly parallel to the coast. These bands of plant and animal life are best seen from the shore when one studies a vertical rocky face at low tide (fig. 14). There are obvious differences between the species of organisms growing at and above the level reached by the water at high tide and those growing in that region delimited by the high and low tide levels. Within these limits there may also be a visible zonation of organisms. This kind of zonation is not confined to the area between the tides but extends out into the diver's world all the way to the limits of plant growth.

The following describes these "bands" of plants and lists some of the more common seaweeds which may be found in each. It is important to remember that depending upon local conditions these areas may be narrow or very broad, and many seaweeds are not restricted to a single "band" but may be found in several of them. Also, the seaweeds named as examples may not be found in all latitudes encompassed by this book.

1. Intertidal Zone.

This refers to that area which is alternately exposed and submerged by the tides. Along our coast, the tides are semi-diurnal, that is, they rise and fall twice a day. World-wide the vertical increase in tidal depth ranges from less than 1 foot to more than 40 feet. The

Figure 14. *Calothrix*, a blue-green alga, grows in a dark band just above the barnacles, high in the intertidal zone. Below the barnacles note *Ascophyllum nodosum*.

Atlantic Coast of the North American continent covers almost all of this range.

Plants growing in this intertidal zone must be adapted to extreme stresses. At various times they may be exposed to the air, sun, rain or snow, and at high tide they are completely submerged. Obviously, different levels within the intertidal zone have different intervals of exposure, and thus within this zone one can readily observe a stratification or "banding" of particular species which have adapted to specific time periods of drying out and inundation.

The uppermost level of the intertidal, which is closest to the shore, is uncovered most of the time. Only the hardiest of organisms survive here, and this level is often clearly marked by a band of microscopic blue-green algae such as *Calothrix* (plate 71) which appears black, and looks almost like a stripe of dark paint on the rocks (fig. 14). Small green tubes of *Enteromorpha* (plates 2, 14) may be seen projecting here and there through the *Calothrix,* and there may be abundant growths of *Porphyra* (plate 1).

Below the blue-green algae is often a mass of barnacles and periwinkles, the former's white shells clearly seen as a distinct white band. Both of these organisms may grow throughout the entire intertidal zone.

The remainder of the intertidal is dominated by the brown seaweeds. Below the barnacles, still in the upper levels of the intertidal, one may find *Fucus spiralis* (plate 21), a relatively small *Fucus* species lacking air bladders. Lower in the intertidal the large *Fucus vesiculosus* (plate 18) takes over, and its blades, rich in air bladders, drape over and cover wide areas of rocky substrate. At the lowest levels, where the plants are exposed for only a few hours each day, may be found *Fucus distichus* (plate 20), easily recognized by its large, flattened reproductive receptacles.

On shores exposed to only moderate wave action, the large olive-

Figure 15. The intertidal zone may be quite rich in seaweed growth.

green straplike fronds of *Ascophyllum nodosum* (plate 22) may grow intermingled with the *Fucus vesiculosus.* If the shoreline is very sheltered, the latter may be absent, and *Ascophyllum* may grow in abundance.

Although the intertidal contains more brown seaweeds in terms of plant mass, many other species grow there as well. The purple-red membranes of *Porphyra* (plate 1) may be seen dangling from rocks or spread out over the large seaweeds beneath. The green *Ulva lactuca* (plate 3) is here too, and the golden brown masses of *Ectocarpus* (plate 63). The rocks and shells beneath the seaweeds may be encrusted with the varying red hues of *Hildenbrandia* (plate 73), or south of Cape Cod the delicate blades of *Petalonia* (plate 9) may be found during the summer months.

2. Sublittoral Zone.

This area is always covered by water, except at its uppermost level which may be very briefly exposed during extreme low tides. The boundary between the lower limits of the intertidal and the upper limits of the sublittoral is often characterized by a dense growth of "Irish moss," *Chondrus crispus* (plate 33) and the closely related species *Gigartina stellata* (plate 34). There may be an underlying cover of the encrusting red algae. All of these may extend well out into the sublittoral.

At a depth of 10 feet or so a dominant feature of the sublittoral begins, the presence of large beds of *Laminaria* (plates 7, 11-13) particularly north of Cape Cod, and the green ropelike strands of *Codium fragile* (plate 39) south of the Cape. *Codium,* in fact, has taken over large areas of the sublittoral predominantly at depths from 3 to 10 feet and is one of the first seaweeds encountered upon entering the water from the shore.

Other seaweeds which form dense growths are *Chorda* (plates 37, 38) and *Corallina* (plate 17). As one gets into deeper water, beyond 30 feet or so, one begins to find such species as the large perforated *Agarum cribrosum* (fig. 16) and smaller, finely branched red algae like *Spermothamnion* (plate 65) or *Plumaria elegans* (plate 29).

Rather than elaborate further here, the descriptions of the illustrated plants will list the depth at which the plants may be found,

Figure 16. ***Agarum cribrosum*** **is restricted to the sublittoral zone and to habitats north of Cape Cod.**

always remembering that the reported depths and ranges may change with further exploration and research.

HOW TO COLLECT AND KEEP SEAWEEDS

At the very beginning, a word of caution should be introduced. In terms of ecology, the very best way to collect seaweeds, or any living organisms for that matter, is to photograph them. An organism taken out of its immediate environment can no longer contribute to that environment, and mankind has had a hand in the extinction of many species because of excessive collection and destruction. The use of this book will hopefully be a stimulus to the discovery, identification and appreciation of the seaweeds, but if such discovery leads to careless and superfluous collecting, much harm can be done. It would not take long to denude an area of plants which have taken years to establish themselves.

However, the urge of the naturalist to pick and keep admirable specimens is within us all, so that the following describes how best to sparingly collect and retain specimens of particular interest. For a diver, a "simple" photograph can be quite expensive, and without highly specialized underwater equipment and proper techniques such pictures are often unsatisfactory. The techniques listed below are simple and utilize readily available and inexpensive materials.

The collector should make every effort to take plants that are already detached. These are very common and can often be found lying about the ocean floor, or for the beachcomber, tossed up on the beach by waves. Specimens which are faded should not be used as they are undergoing the process of decay. Degenerating red algae may appear orange, and the brown and green algae become whitened, particularly when exposed to the sun on the shore.

When picking seaweeds, one representative plant of each species is sufficient. Care should be taken to include the holdfast which may be removed from the substrate with a knife, and all specimens should be kept damp and cool until further treatment. After removal from the collecting bag, they should be placed in plastic bags and kept in a cool dark container such as a Styrofoam cooler. Seaweeds degenerate quickly, and after a few days, even under these conditions, they will no longer be suitable.

The ideal way to keep and display seaweeds is to press and dry them on stiff sheets of paper. They can also be maintained indefinitely in liquid preservative, but unless one is interested in maintaining them for scientific purposes, such preservation is unnecessary, not to mention aesthetically unpleasant. Dry pressed seaweeds, particularly the red algae, make very beautiful displays and, properly matted and framed, will keep their color and form for many years.

For those interested in a simple liquid preservation technique, a

5% solution of formalin in sea water will suffice. In pressing specimens, the following equipment is useful.

1. Equipment.

a large plastic cafeteria tray *or* a large enamelware tray
herbarium paper (100% rag is best)
blotting paper
corrugated cardboard
wax paper (or old sheets or cheesecloth)
bulb syringe (a large meat baster)
needles or a camel hair brush

First, an explanation about the above equipment. The trays must be plastic or enamelware to avoid corrosion from the sea water. The herbarium paper is a standard 11-1/2" x 16-1/2" paper used by botanists to press plants for scientific reference, the 100% rag type being best because it is a strong, absorbent paper. To economize, it is not necessary to use an entire sheet of paper for each plant, as smaller specimens can be pressed onto halved or quartered sections. A heavy blotting paper, also 11-1/2" x 16-1/2", is best to efficiently absorb excess moisture but folded newspaper will suffice. A corrugated cardboard allows air to flow along the cardboard surface and past the blotters for quick drying. The use of cardboards is quite helpful, but not essential. The wax paper or a substitute is used to prevent the seaweed from sticking to the blotter. Wax paper is the most convenient material, old sheets cut into pieces are better if available, and cheesecloth is good but expensive. The syringe and needles are used to arrange the plant neatly on the paper before pressing. The herbarium paper, blotting paper and cardboard may be obtained from a biological supply company, such as Carolina Biological Supply, Burlington, North Carolina 27215 or Connecticut Valley Biological Supply, Valley Road, Southampton, Massachusetts 01073.

2. Procedure.

The wet plant is placed on the herbarium paper, which has been moistened with clean sea water and placed on the tray. The plant is then carefully arranged in the desired position by a gentle stream of sea water from the bulb syringe, and/or by manipulation with the needle or camel hair brush. Sometimes the plant is too large or bulky to fit properly on the paper. If the plant is too "bushy" pieces may be cut off to flatten it, or if the holdfast or stipe is too thick, it should be trimmed along the back so that its front surface remains on the paper. Very long specimens such as *Chorda* or *Laminaria* can be coiled or folded. Care must be taken to clean the paper of any debris or fragments which will spoil the ultimate appearance of the specimen.

The tray is then gently tilted to allow excess water to run off. The paper is slid off the tray onto a blotter, which has been placed on the corrugated side of a piece of cardboard. Then a piece of wax paper or substitute (see above) is placed on top of the plant and paper. On top of this is placed another blotter, and then another cardboard, corrugated side down. All of this represents one "unit" of the press. Such "units" or "sandwiches" of plants are piled up to a height of 18-24 inches, and should then be subjected to pressure. This can be applied simply by placing heavy stones or books on the pile, or by using a standard plant press, also commercially available. The pressed piles should be kept in a warm, dry place and the blotters should be changed at least once a day. The secret to proper drying is to allow the specimens to dry at a moderate rate, which is brought about mainly by changing the blotters. Sometimes a small fan to circulate the air through the pile is helpful, but excessive heat such as direct sunlight should be avoided. The average drying time is about two days for the smaller seaweeds to one week for the largest.

If the specimens are removed before they are completely dry, the plant will continue to dry and shrink, and the paper will buckle up and wrinkle. It is a good idea to partially dry some of the larger brown algae such as *Laminaria* (plates 7, 11-13) or *Agarum* (plate 4) between folded newspapers before pressing. They otherwise might become moldy or decayed before they are completely dry.

The seaweeds have a slimy outer layer which acts as a natural glue and most species will adhere quite firmly to the paper. A few stiffer ones like *Ahnfeltia* (plate 44) or *Ascophyllum* (plate 22) do not, and after drying should be fastened to the paper with small strips of white cloth tape.

Fucus species (plates 18, 20, 21) present another problem, in that the reproductive receptacles often break during pressing and stain the paper. This can be avoided by pressing them between sheets of newspaper first, and when they are dry taping them to the herbarium paper. Also, fresh water should never be used to soak or moisten the plants, as it may rupture the cells and ruin the appearance of the specimens.

One need not press the calcified specimens such as *Corallina* (plate 17) or *Lithothamnion* (plate 74). Simply collect small rocks (sparingly) with these organisms growing on them, rinse, and dry them. They will maintain their appearance indefinitely.

The preservation of the larger specimens of *Laminaria* (plates 7, 11-13) presents a special problem to the collector. These can be kept very nicely by first soaking them until saturated in a solution of 3 parts glycerine, 3 parts alcohol, 3 parts sea water and 1 part carbolic acid (phenol). They are then drained by hanging for several hours, and after soaking them for several days in pure glycerine and draining again, they will have developed a moist, flexible texture that can be maintained by storing them rolled in a plastic bag. They can be removed from the bag for display or study for hours at a time without drying out. This technique can be used with other genera such as *Alaria* (plate 5), *Agarum* (plate 4), *Ascophyllum* (plate 22) and *Fucus* (plates 18, 20, 21).

USEFUL AND EDIBLE SEAWEEDS

Mankind has been utilizing the seaweeds in many ways for centuries, including using them as food sources, by either eating them directly or by eating products derived from them. However, since the seaweeds are generally low in protein content and are made up principally of indigestible cellulose, the major benefit of such consumption is from their salt, vitamin and mineral content. Such limitations have not deterred widespread use of these plants, particularly in the Orient, where seaweed harvesting has been known for 5000 years. Our culture has for the most part lost its interest in seaweeds as delicacies, but in the Orient many species of seaweeds are eaten and enjoyed daily by millions. The novice "algae-eater" may not find all seaweed preparations to his liking, but experimentation with some of our common local species will result in the discovery of some delicious additions to the menu. Listed below are some species, most of which are within the range covered by this book, which may be eaten and in some cases have very important industrial uses. Descriptions of such uses and recipes for the preparation of these species are included. Several commercially valuable seaweed extracts are also listed.

***Porphyra* (*plate 1*).**

Also called "nori" or "laver," *Porphyra* is so popular in Japan that it is actually farmed in the quiet, shallow bays along the coast. Long bamboo or fiber rafts, from which hang strings of oyster shells, are put out in November to pick up the reproductive cells of *Porphyra* and the resulting plants are harvested by hand from December to March by workers who float along next to their "crops" and pick as the tide goes out.

Such cultivation has been going on since 1670, and now the annual harvest is over 11,000 dry tons! The membranous plants are washed and dried and sold as packets of flat, dry sheets. These plants are eaten in a variety of ways, most commonly as "Sushi"—wrapped around boiled rice, meat, fish or eggs. It can be toasted over a charcoal fire and broken up and added to soups and sauces. It may simply be dropped in soy sauce and eaten by itself. An easy and very tasty preparation is dried *Porphyra* sautéed in peanut or olive oil. When collecting it, let it dry overnight by hanging over string or wire, and before adding it to soup or making "Sushi" boil it for a minute or so in fresh water.

***Palmaria palmata* (*fig. 17*).**

Commonly known as "dulse," this conspicuous red seaweed has been eaten for twelve centuries in Europe. In northern Maine and the Canadian Maritimes it is collected, dried and sold to be eaten either directly or as an addition to soups, sauces, salads or relishes to which, in small amounts, it imparts a delicate salty flavor. It is often seen in "health food" stores along with other seaweed products but is really of limited nutritional value. It is said that eating dulse the first few times can be unpleasant (which the author can verify) but that the taste for it soon develops (which the author cannot verify). At any rate, it can be easily collected and hung out to dry in the same manner as *Porphyra*.

Figure 17. The edible *Palmaria palmata* grows abundantly in cold water. Here it is attached to a *Laminaria* stipe.

***Chondrus crispus* (*fig. 18*).**

Very rich growths of *Chondrus crispus* or "Irish moss" occur along the coasts of Maine and the Maritime Provinces. It develops as a dense covering on the rocks up to a depth of about 70 feet and is a conspicuous part of the sublittoral vegetation from Cape Cod to Newfoundland. About 600 years ago, housewives in County Carragheen, Ireland, discovered that a handful of this seaweed cooked with milk and flavored made a delicious pudding. Many other uses for the plant developed, and it was utilized for centuries in Ireland and western Europe as food, medicine and fertilizer.

Figure 18. The highly valuable *Chondrus crispus*, more commonly known as "Irish moss."

When the Irish immigrants first came to this country, they imported their Irish moss until in 1835 a former mayor of Boston, Dr. J.V.C. Smith, discovered that it grew locally and this stimulated an industry involved in its collection and preparation.

When *Chondrus* is dried and boiled in water a gummy substance is extracted, known chemically as a hydrocolloid. This is made up of very large molecules that have the property of remaining dispersed and suspended in liquids without settling to the bottom. In other words, it can "thicken" water. Such properties have enormous potential for use in the food industry, and the evidence of this potential can be demonstrated by the fact that at the world's largest seaweed processing plant, Marine Colloids, Inc., at Rockland, Maine, over 60,000 pounds of seaweed are extracted daily. Almost half of the powdery extract called "carrageenan" comes from *Chondrus,* most of which is collected along the Maritimes. Less than 5% is collected from Massachusetts, Maine and New Hampshire. There, from May to November, collectors wield long-handled iron rakes from their boats and break off the plants from their holdfasts which cling tightly to the rocks. New plants can then regenerate from these holdfasts. This crude technique has proven more efficient than any other, but it is likely that divers could at least be of assistance in locating *Chondrus* beds which could then be worked over.

The *Chondrus* plants are dried, bundled and shipped to Rockland or to another large plant in Portland, Maine, operated by Kraft Foods. It has recently been estimated that the current yield of *Chondrus* from the Maine coast could be safely boosted to an annual harvest of 30-50 million pounds.

Carrageenan is extraordinarily versatile and may be used as a stabilizer and texturizer in milk products (chocolate milk, evaporated milk, ice cream, etc.) and in puddings, pie fillings and whipped toppings. It is used as a jelling agent in dietetic jellies, and in non-dairy creamers. It acts as a binder in toothpastes and hand lotions. It is used in syrups, sauces and frozen food concentrates. A new division of Marine Colloids, Inc., M.C.I.B. Biomedical, is now researching and marketing clinical diagnostic products based on the chemical properties of seaweed hydrocolloids.

Try making some Irish moss pudding. Dry some of the seaweed, and add together the proportions of 1/4 cup dried *Chondrus* to 1 pint milk. Cook in double boiler for 1/2 hour, add sugar and flavoring to taste, strain and cool. It makes a delicious pudding. You may even want to try this one—Irish bartenders in some New York pubs soak Irish moss in whiskey and offer it as a cough cure.

Laminaria (fig. 19).

No one who dives or just walks along our shores can fail to notice the large brown blades of *Laminaria,* either growing in dense, waving masses in the sublittoral or detached from the stipe after a storm and tossed up on the beach. Few, however, would consider eating this plant, but in Japan the *Laminaria* stipe is a foodstuff. In fact the artificial sweetener sodium glutamate was discovered from *Laminaria* in Japan in 1908 by Dr. K. Ikeda.

The stipe is cut into strips, softened by soaking in vinegar, and either chopped up and boiled with sugar as a tea substitute or eaten as a kind of pickle. These make very tasty additions to salads or sandwiches and can be adapted for use in many relish recipes. Konbu, the Japanese word for *Laminaria,* means "delight." In Russia, *Laminaria* is prepared by first washing and drying it, and after soaking in fresh water for three to four hours it is chopped into small pieces and fried in vegetable oil for five minutes, and quickly

Figure 19. ***Laminaria saccharina,*** **the common seaweed whose stipe is eaten pickled or cooked. It grows in dense sublittoral beds.**

cooled. To this is added a mixture of cabbage, beets and tomato sauce.

Alaria esculenta (*plate 5*).

The word *esculenta* means "tasty" and both the flattened midrib and the small reproductive blades growing from the stipe can be eaten. They can either be washed, chopped up and added directly to a salad, or can be dried and crumbled and used as a salty seasoning.

***Enteromorpha* (plates 2, 14).**

Known as "Anori" in Japan, these common bright green membranous tubes can be dried in bundles and then broken up or crushed into powder and eaten seasoned with soy sauce, or used as a salad ingredient.

***Codium fragile* (plate 39).**

Considering the way in which *Codium* has bloomed in the waters of Long Island Sound and southern Cape Cod it might be desirable to have this plant considered a delicacy. The Japanese prepare it by bleaching it in the sun, and then soak it in a sugar solution before eating. They call it "Miru."

Kelp.

This term originally referred to the ash derived from burning seaweeds. It has come to refer to the large brown seaweeds, particularly the West Coast genera *Macrocystis* and *Nereocystis,* but can include local plants such as *Laminaria* (plates 7, 11-13) and *Agarum* (plate 4).

Kelp processing began in France in the 17th century and in this country during World War I there were ten factories in California burning and processing seaweeds to extract materials such as iodine, potassium chloride, acetone and potash. Presently other cheaper sources for these materials were found, but the kelp industry expanded into the production of algin. This is a chemical extract, also a hydrocolloid, prepared from large brown seaweeds collected off the California coast. After a complex process, the algin is isolated from the cell walls of these plants and modified into soluble derivatives such as calcium or sodium alginate. These dissolve in water to make thick, syrupy solutions which do not coagulate when heated or jell when cooled. Algin derivatives have an impressive list of uses in the food, pharmaceutical, rubber, dairy and textile industries.

For example, they give ice cream a smooth texture by preventing the formation of ice or sugar crystals and by separation of fat globules. They suspend the abrasive in auto polishes and the pigment in paints, and act as an emulsifier in such diverse products as cosmetics and French dressings.

Off the California Coast, enormous collecting barges pass over the seaweed beds and cut off only the upper 4 feet of the plants so that new growth will supply another crop. In this manner they may collect as much as 300 tons per day.

Agar.

This is an extract from red seaweeds, principally *Gelidium* and *Gracilaria* (plate 26). The latter grows along our coast, but is not collected for commercial use around here. *Gelidium* is hand picked by divers in the warm waters of Japan, China, California and Mexico.

The agar industry was started in Japan in the middle of the 17th century and there are presently numerous processing plants in Japan which are engaged in agar extraction. When dissolved in hot water, agar forms a solution which upon cooling develops into a firm, jelly-like substance. It is eaten in the Orient as a dessert base but like algin and carrageenan has many other uses as well. It is particularly useful in foods where firm jels are required such as jellied fowl, fish and meats, or in chiffon pies, meringues and icings. It is also used in photographic films, dental impression materials, and many bacteriological and medical preparations. Probably its most well-known use is in the preparation of agar plates of growth medium for bacteria in clinical and research studies.

Fertilizer.

Seaweeds have been used for centuries as fertilizer, particularly in coastal Asia and Europe. It has the advantage of lacking weed seeds and spores of crop diseases, and although somewhat low in nitrogen and phosphorus, it is rich in potassium and other minerals. If you're interested in "organic" farming, collect piles of seaweeds that have been tossed up on the beach and add them to your compost pile. They break down slowly but are eventually good soil builders.

SEAWEED RESEARCH WITH S.C.U.B.A.

A textbook on the algae published in 1965 pointed out that the seaweeds of the sublittoral zone had received relatively little attention up to that time except from examination of dredged material. Dredging is helpful in locations which exceed safe diving depths, but give an altogether incomplete picture of the bottom flora which should be supplemented by the observations of divers where this is practical. In other words, divers in the oceans have the opportunity to chart unexplored territories.

Since that time, S.C.U.B.A. has emerged as a valuable tool in research concerned with the ecology and distribution of the seaweeds. For example, Dr. Robert Wilce of the University of Massachusetts and his students have done extensive diving in the Arctic, and have furthered our knowledge of deep, cold water seaweeds. In 1967-1968 Dr. Martine (Villalard) Bohnsack of Roger Williams College in Rhode Island studied the interrelationships of 104 species of seaweeds off Conanicut Island, and in 1966-1969, in the most extensive S.C.U.B.A. research done to that date in New England waters, Dr. James Sears of the University of Massachusetts conducted a comprehensive study of the sublittoral seaweeds of southern Cape Cod and adjacent islands. He found 142 species, 23 of which had never before been reported for Cape Cod!

Others have also reported new ranges of seaweeds discovered while diving. In 1967 Edelstein and others from the Canadian National Research Council first reported *Alaria grandifolia* in Nova Scotia at depths of 40-60 feet. Dr. Arthur Mathieson of the University of New Hampshire discovered *Halicystis,* never before reported from the Northeast, when diving at 40-80 feet.

By 1986, numerous botanists were using S.C.U.B.A. as a routine method for seaweed sampling and direct observation of underwater plant life, and some had also begun to employ manned submersibles such as the Florida Harbor Branch Foundation's Johnson-Sea-Link I.

Thanks to careful observation by S.C.U.B.A. diving scientists in California, the alarming decline in the giant kelp beds harvested for the production of algin was found to be the result of grazing sea urchins which had experienced a population explosion due to the gradual disappearance of their natural predator, the sea otter. They found that addition of quicklime to the water killed off the sea urchins and this technique coupled with a resurgence of the sea otters in some areas has saved the kelp industry.

It is an obvious advantage to be able to see plants exactly as they occur in nature, and to note precisely where they are growing, how abundantly, and with what other plants they are associated. Even most botanists, for example, ordinarily see a plant such as *Ahnfeltia plicata* (plate 44) as a bleached fragment washed up on the beach, whereas the diver may see it at a depth of 50 feet, as a deep red or black bushy plant.

S.C.U.B.A. is therefore not only recreational but may be quite useful in scientific research. With a little study and effort perhaps the reader will be able to assist in expanding our knowledge of life under the sea.

KEY TO UNBRANCHED BLADES OR MEMBRANES

1. A purple or brown very thin membrane	*Porphyra* (plate 1)
1. Not as above	2
2. Bright green	3
2. Not as above	4
3. A narrow, wavy green blade, tubular at base	*Enteromorpha linza* (plate 2)
3. A green, very thin wider membrane	*Ulva lactuca* (plate 3)
4. Blade or membrane with midrib	5
4. Not as above	7
5. Blades perforated with many holes	*Agarum cribrosum* (plate 4)
5. Not as above	6
6. Long brown blade, small blades along stipe	*Alaria esculenta* (plate 5)
6. Red, delicate membrane	*Grinnellia americana* (plate 6)
7. Blades smooth, leathery, long (up to 1 ft. or more)	8
7. Blades smaller, may be coarse and hairy	9
8. Blade divided into several to many straplike segments	*Laminaria digitata* (plate 7)
8. Blades solitary	10
9. Small brown blades, up to 18 inches long	*Punctaria* sp. or *Petalonia* sp. (plates 8, 9)
9. Larger, coarse blade, many hairy tufts	*Sacchorhiza dermatodea* (plate 10)
10. Stipe very long, often inflated at upper end	*Laminaria longicruris* (plate 12)
10. Not as above	11
11. Stipe shorter	*Laminaria saccharina* (plate 11)
11. Stipe very short, spade-like blade	*Laminaria platymeris* (plate 13)

Porphyra umbilicalis

Porphyra umbilicalis **pour - fý ra um - bill - lee - káy - liss**
porphyra: purple **umbilicalis: of the naval.**
Common Names: "purple laver," "nori"

The genus *Porphyra* is easy to recognize. It is a very thin, large membrane with a distinct satiny sheen. It is attached by a very small holdfast to other seaweeds and rocks in the intertidal and shallow sublittoral zones, often forming dense colonies. In order to positively determine the species, one must examine sections of the specimen in question under the microscope, but several more convenient diagnostic features should be utilized. The common *Porphyra umbilicalis,* here illustrated, is olive or brown-purple and may be more than 1 foot in diameter. The term *umbilicalis* refers to the fact that the small holdfast tends to be centrally located and the membrane tends to have a pinched appearance in this area.

The smaller *Porphyra leucosticta* is dull pink to reddish purple and its range extends as far south as North Carolina, whereas the southern limit of growth of *Porphyra umbilicalis* is New Jersey. *Porphyra miniata* may also be up to one foot in diameter but is a definite rosy color and is restricted to habitats north of Cape Cod.

Porphyra undergoes a complicated reproductive process which, like that of many other seaweeds, is not completely understood. The genus *Conchocelis,* a red, filamentous shell penetrating plant has been shown to be, in fact, one of the stages in the life history of *Porphyra,* so that it is at one time a membrane, and at another a filament.

Please consult the Introduction (p. 34) to find out how this commercially valuable seaweed can be prepared for consumption in a variety of ways.

Similar Species: *Ulva lactuca* (plate 3) is membranous but always bright green.

Range: Virginia to the Arctic.

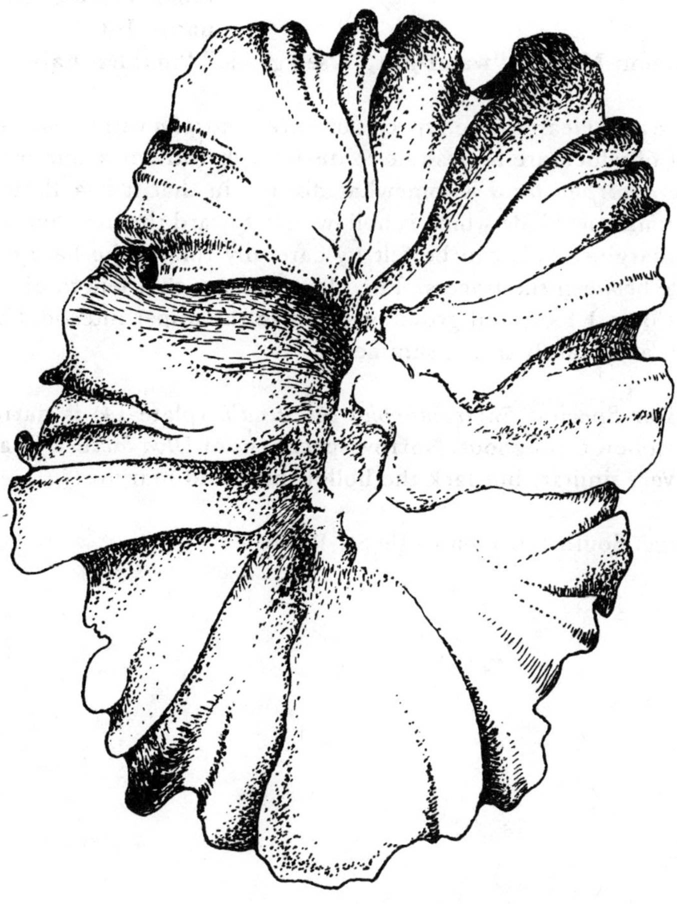

Plate 1. Porphyra umbilicalis

Enteromorpha linza

Enteromorpha linza
enteromorpha: intestine-shaped

én - te - ro - mór - fa lín - za
linza: after Linnaeus, a famous 18th century naturalist

Common Names: "water-gut," "sea grass," "maiden hair"

There are at least a dozen species of *Enteromorpha* within our range, most of which are impossible to distinguish without a microscope. *Enteromorpha linza* is somewhat distinct in that it is a flattened, green narrow blade which is hollow only toward the base and along the margins. This can be felt by carefully rolling the base of the blade between the fingers. The plant may reach a length of somewhat over 1 foot, and grows on rocks in the upper intertidal zone, often abundantly in the summer.

Similar Species: *Enteromorpha intestinalis* (plate 14) is narrower and hollow throughout. Narrow specimens of *Ulva lactuca* (plate 3) are very similar, but lack the hollow base and margins.

Range: South Carolina to Ile St. Pierre.

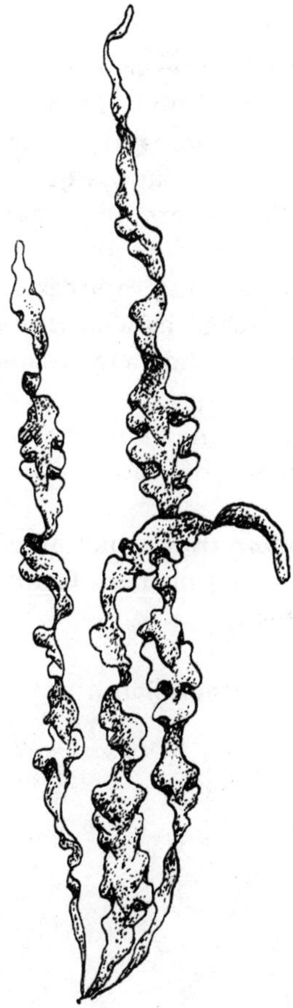

Plate 2. Enteromorpha linza

Ulva lactuca

Ulva lactuca **úll - va lac - tóo - ka**
Ulva: water **lactuca: lettuce**
Common Names: "sea lettuce," "green laver"

The bright green membranes of *Ulva lactuca* are a common sight along our shores all year round. These green sheets, only 2 cells in thickness, may be 2 feet long and almost as broad, but are usually considerably smaller. The shape is quite variable, some specimens being almost circular, others quite narrow and elongated. The plants have a fine, silky texture and the margins are waved or ruffled. *Ulva* may be very dense in areas such as quiet bays and salt marshes. It grows attached by a small disc but is very often found floating in a rather torn and ragged condition. It retains its color very well after being pressed.

Ulva is edible in the same manner as *Porphyra* (p. 34) but is not considered as much of a delicacy.

Similar Species: *Monostroma* (not illustrated) is very much like *Ulva* but is only 1 cell in thickness. Confusion between these species is therefore inevitable.

Range: Florida to Newfoundland.

Plate 3. Ulva lactuca

Agarum cribrosum

Agarum cribrosum **a - gá - rum crib - rów - sum**
agarum: a fungus name **cribrosum: full of sieves**
Common Names: "sea colander," "devil's apron"

Agarum is another one of those seaweeds generally available in good condition only to divers. It grows in cold water at a depth of around 30 feet and deeper, and plants which wash up on shore usually are fragmented and have lost their holdfast. The stipe is short, and supports a unique, heavily perforated blade which is divided by a flattened midrib. The blade is usually somewhat ruffled and is quite leathery and dark brown to olive in color. Dwarfed forms may occasionally be found in deep tide pools, but the plant is ordinarily about 1½ feet long and 1 foot wide.

Similar Species: The perforated blade serves to distinguish *Agarum* from all of the *Laminaria* species, to which it is closely related (plates 7, 11-13).

Range: Northern Massachusetts to Labrador.

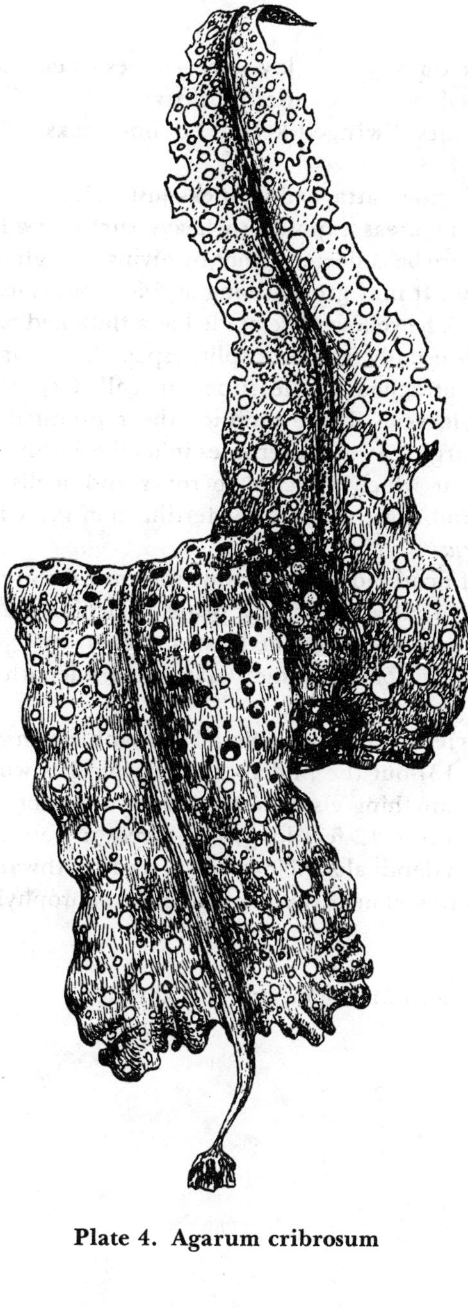

Plate 4. Agarum cribrosum

Alaria esculenta

Alaria esculenta **a - láy - ree - a es - cue - lén - ta**
alaria: winged **esculenta: tasty**
Common Names: "winged kelp," "dabberlocks," "henware"

This seaweed grows attached to rocks just below low tide out to at least 25 feet in areas exposed to heavy surf or swift currents. It should therefore be collected either by diving at high tide or by wading at low tide. It may grow as long as 6 feet but is usually half this length or less. The yellow-olive blade has a flattened midrib along its length, and is usually frayed or split, especially toward the tip. The small blades attached to the stipe are called sporophylls, which means "seed-leaves." These produce the reproductive cells of the plant. This large seaweed alternates in its life history with a microscopic filament which attaches to rocks and shells and produces sperm cells and eggs, which after fertilization grow into the larger form of *Alaria*.

In parts of Scotland and Ireland, the midribs are removed and served as one would serve string beans. There are about 5 species of *Alaria* in our area, but at least 7 others occur on the North American Coast, some of which may reach a length of 20 feet.

Similar Species: *Alaria* superficially resembles *Laminaria* species (plates 7, 11-13) but the midrib and sporophylls will easily distinguish it from anything else. *Alaria grandifolia* (not illustrated) is a larger plant, up to 12 feet long, and grows in cold water out to at least 60 feet in depth along Nova Scotia and northward. The stipe is somewhat flattened near the blade and the sporophylls are up to 16 inches long.

Range: Northern Massachusetts to Labrador.

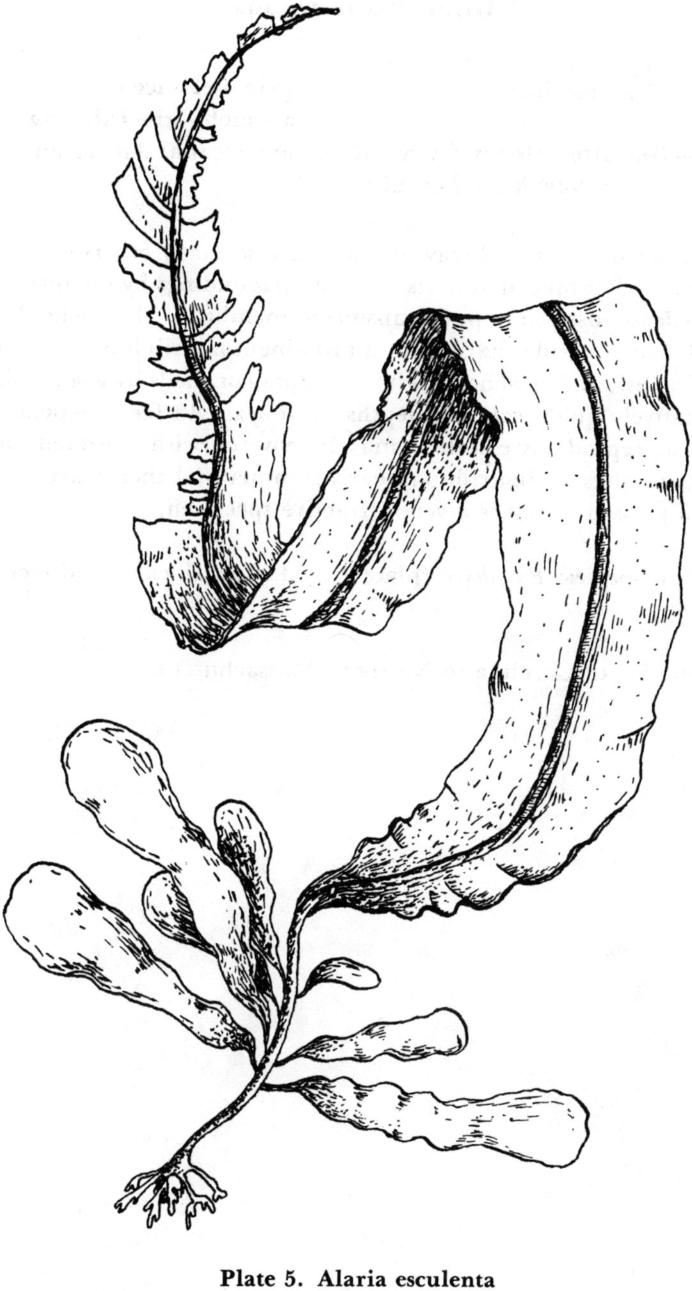

Plate 5. Alaria esculenta

Grinnellia americana

Grinnellia americana — **grin - él - lee - a**
a - méh - ri - cáh - na

grinnellia: after Henry Grinnell, a New York botanist — **americana: American**

This is a truly beautiful seaweed, and like so many others can only be seen and appreciated in its natural surroundings by the diver. It is a delicate rose-red or pink translucent membrane, 4-8 inches long and 1-4 inches wide. Except for a prominent midrib it is only 1 cell in thickness, and attaches to wharves, stones or shells by a small disc in relatively quiet water at depths to at least 40 feet. Depending upon its reproductive state it may be covered with scattered dark granular spots. It flourishes in early summer and then disappears. When pressed it makes a very attractive specimen.

Similar Species: *Porphyra* (plate 1) is darker, thicker, and lacks a midrib.

Range: South Carolina to Northern Massachusetts.

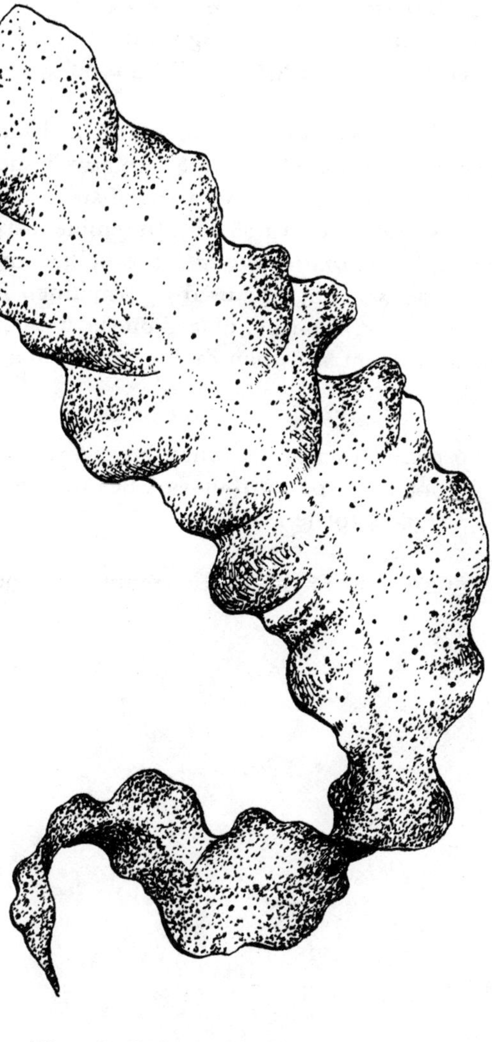

Plate 6. Grinnellia americana

Laminaria digitata

Laminaria digitata **lam - in - á - ree - a dij - i - táy - ta**
laminaria: a thin plate **digitata: fingered**
Common Names: "fingered kelp," "tangle"

The blade of this species may be much divided into as many as 30 slimy brown segments, and the stipe is relatively short and flattened toward the blade. It flourishes in areas of swift, cold currents from just below low tide to at least 65 feet. In remote districts of England, whole plants of *Laminaria digitata* are boiled and used as cattle food. Years ago, its stipe was used by fishermen for making knife handles. Sections of the stipe, placed on the base of the knife blade while still fresh, contract upon drying and become solid. Try it, it really works.

Similar Species: Other species may have divided blades, but are most likely not true species, but varieties of *digitata*, for example, *Laminaria platymeris* (plate 13).

Range: Long Island to the Arctic, more common north of Cape Cod.

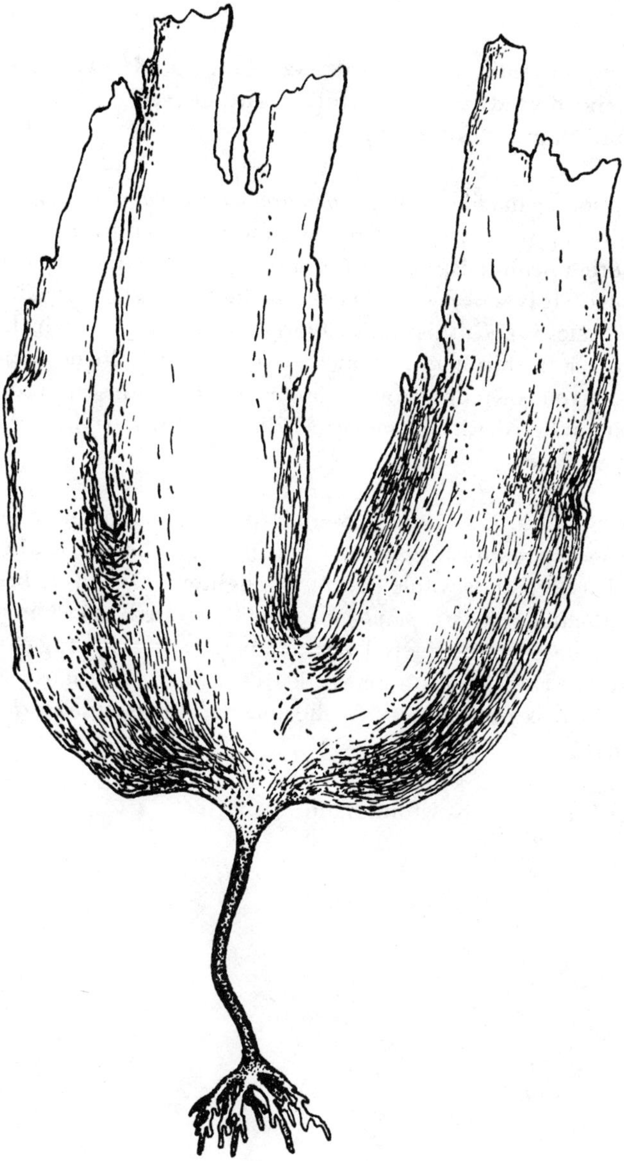

Plate 7. Laminaria digitata

Punctaria latifolia

Punctaria latifolia **punk - táy - ree - a la - ti - fó - lee - a**
punctaria: dotted **latifolia: wide-leaved**
Common Name: "dotted weed"

Looking somewhat like a small *Laminaria saccharina* (plate 11), *Punctaria* can be found in early spring and summer attached by a small disc to stems, other seaweeds such as *Chondrus* (plate 33) and is particularly common on leaves of *Zostera* (eel grass), out to at least 40 feet in depth. In the species *latifolia* this disc supports a wiry short stalk which broadens into a soft, thin blade 1-2 inches wide and 4-18 inches long. This blade is olive-brown or yellowish, and is covered with scattered tufts of fine hairs. When dry it becomes greenish and does not adhere readily to the paper.

Similar Species: *Laminaria saccharina* (plate 11) has a smoother blade and a rootlike holdfast. *Petalonia* (plate 9) is dark brown, leathery, and tapered at both ends, while *Punctaria* is lighter and tends to be ragged toward the tip. *Punctaria plantaginea* (not illustrated) is a wider, coarser plant usually heavily covered with filaments and quite ragged. Occasionally one may find, on *Zostera* leaves, a very small narrow brown blade which is not slippery to the touch. This is *Desmotrichum* (not illustrated).

Range: Virginia to Newfoundland.

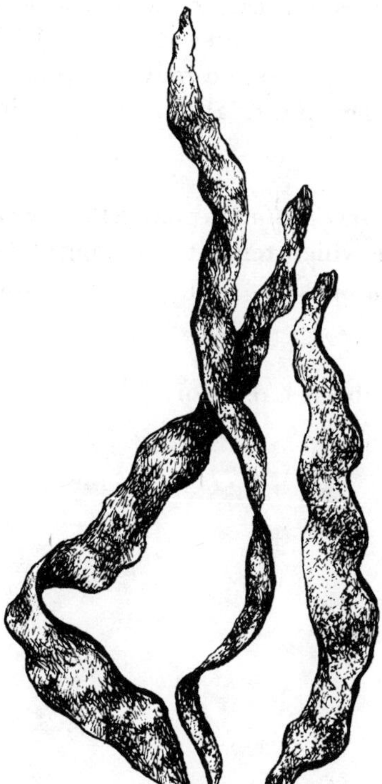

Plate 8. Punctaria latifolia

Petalonia fascia

Petalonia fascia	**pét - a - lów - nee - a fá - sha**
petalonia: a leaf	**fascia: a bundle**

Several tapered blades commonly arise from a small disc-like holdfast which usually attaches to rocks or shells, and less frequently to larger seaweeds such as *Fucus* (plates 18, 20, 21). The blades are coarse, leathery and dark brown and may be up to 18 inches long. They grow from the upper intertidal zone out to a depth of at least 10 feet.

Similar Species: *Punctaria latifolia* (plate 8) has a smoother, thinner, olive brown blade which tends to be ragged toward the tip. Young *Laminaria saccharina* (plate 11) has a smoother blade and a rootlike holdfast.

Range: Florida to Northern Labrador.

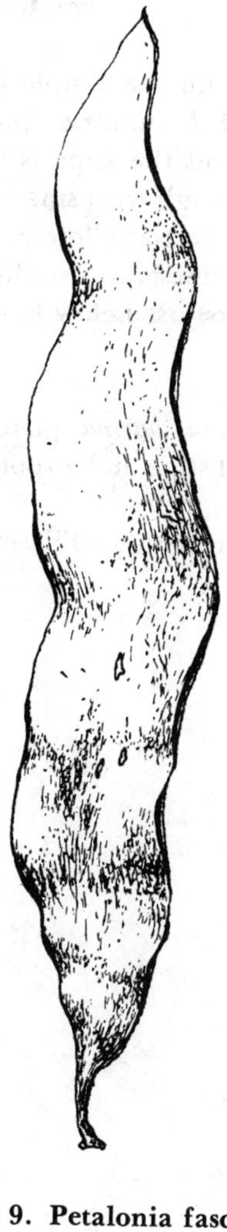

Plate 9. Petalonia fascia

Saccorhiza dermatodea

Saccorhiza dermatodea — **sack - o - ríse - a dér - ma - tóe - dee - a**

saccorhiza: sac root — **dermatodea: skinlike**

This seaweed is patterned on the simple holdfast, stipe and blade pattern similar to that of *Laminaria* (plate 11) and others. The holdfast is a small disc, and the stipe is somewhat flattened. The blade is quite coarse and tough with small hairy tufts scattered over its surface. This dark brown to yellow-brown plant may be 8 feet long and 2 feet wide but does not usually attain such a size. It is found on surf beaten rocks just below low tide, and out to at least 165 feet.

Similar Species: *Laminaria saccharina* (plate 11) lacks the hairs on the blade, and has a cylindrical stipe and a rootlike holdfast.

Range: Northern Massachusetts to Ellesmere Island.

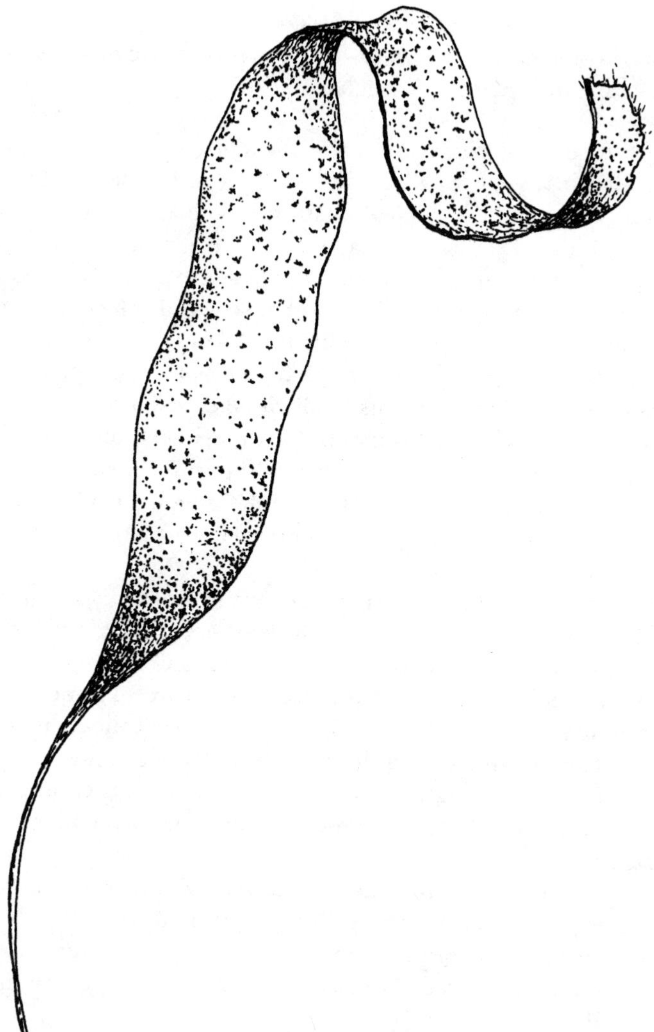

Plate 10. Saccorhiza dermatodea

Laminaria saccharina

Laminaria saccharina	**lam - in - á - ree - a sac - a - ryé - na**
laminaria: a thin plate	**saccharina: sugary**

There are several species of *Laminaria* along our North American Coast. *Laminaria saccharina* attaches to stones and shells from the shallow sublittoral to at least 60 feet. A strong rootlike holdfast supports a flexible cylindrical stipe and a blade which is commonly 6 feet or more in length. The margins of the mature blade are quite thin and often very ruffled. As in the case of many other seaweeds, the large conspicuous plant is just one phase in the life history of the organism, and the male and female reproductive phases of *Laminaria* are represented by microscopic branched filaments totally unlike the large coarse blades.

When dry, this seaweed is covered with a sweet white substance—hence the name *saccharina* which means "sugary." Nineteenth-century writers tell of the practice in England of hanging the blade as a rough barometer—in fair weather it is dry and stiff and if rain is threatening it softens and becomes limp.

All divers within the geographical range of *Laminaria* cannot fail to notice dense masses of these long leathery blades moving back and forth in the shallow to deep sublittoral. Such underwater forests make excellent hiding places for lobsters and careful removal of old blades lying along the ocean floor will often reveal them. It is probably best to avoid swimming down among dense strands of *Laminaria* because even though the blades are slippery and entanglement probably does not often occur, once caught the diver must cut his way out.

Recent studies have indicated that *Laminaria* is an incredibly productive plant. The actively growing portion of the plant is at the junction of the blade and the stipe, and the far end degenerates and falls off as it ages. In cold Canadian waters the entire blade may replace itself at least five times a year.

Similar Species: *Laminaria longicruris* (plate 12).

Range: New Jersey to Newfoundland.

Plate 11. Laminaria saccharina

Laminaria longicruris

Laminaria longicruris **lam - in - á - ree - a lon - ji - crú - riss**
laminaria: a thin plate **longicruris: long-legged**
Common Name: "kelp"

As the name suggests, this species is distinguished by a very long stipe which may reach a length of 6 feet and sometimes be longer than the blade itself. This stipe becomes somewhat swollen and hollow before it joins the blade, and often in shallower water is conspicuously bent and narrowed at the base of the blade. The latter may be 6-25 feet long and quite wide. *Laminaria longicruris* is a cold water plant which grows from the shallow sublittoral to deep water.

Similar Species: *Laminaria saccharina* (plate 11). Some authorities think that *Laminaria longicruris* is just a variety of *Laminaria saccharina* and not a true species.

Range: Northern Massachusetts to Ellesmere Island, occasionally south.

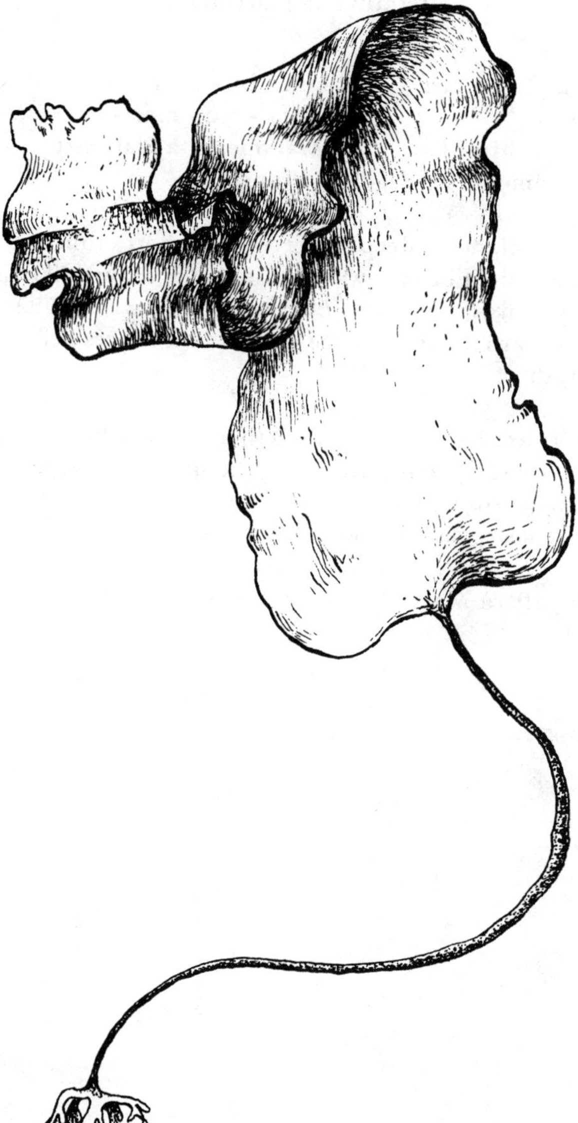

Plate 12. Laminaria longicruris

Laminaria platymeris

Laminaria platymeris **lam - in - á - ree - a**
pla - tee - mér - iss

laminaria: a thin plate **platymeris: a flat part**
Common Names: "sea girdle," "tangle"

If you find a plant in deep water with a short, thin stipe and a broad, dark to light brown, wide blade that is shaped like the spade in a deck of cards, you have *Laminaria platymeris.* This blade is often split toward the tip and the whole plant ranges between 1 to 6 feet in length.

Similar Species: *Laminaria digitata* (plate 7) may have forms which have only several segments, but lacks the spadelike shape. Actually, *Laminaria platymeris* is now regarded by most authorities as a variety of *Laminaria digitata* (plate 7) and not a true species.

Range: Northern Massachusetts to Newfoundland.

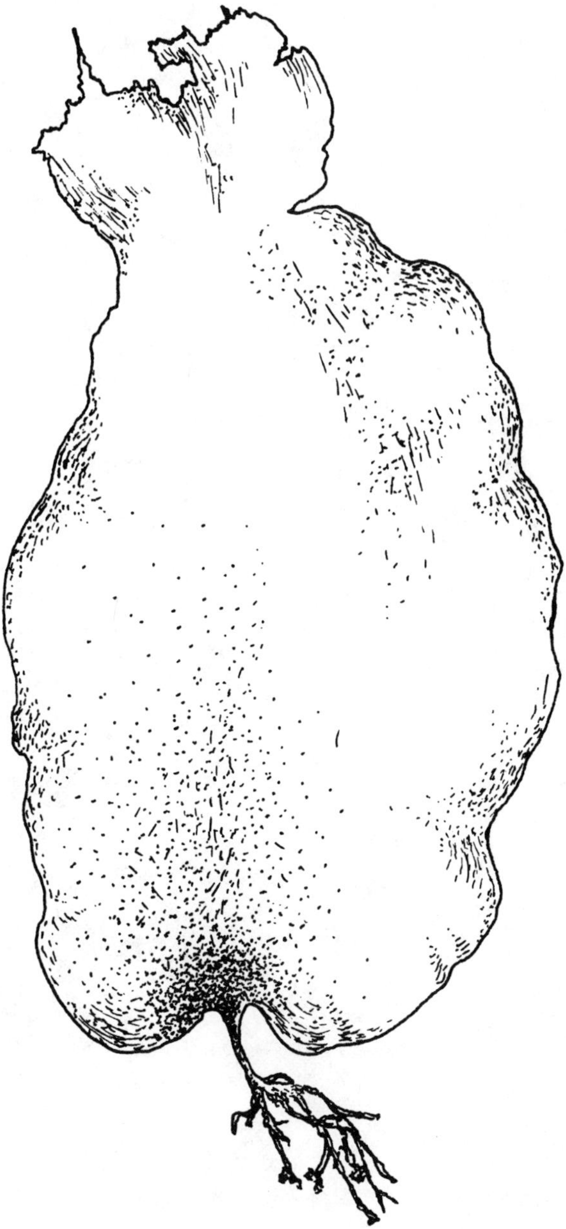

Plate 13. Laminaria platymeris

KEY TO UNBRANCHED TUBES

1. Bright green	*Enteromorpha intestinalis* (plate 14)
1. Not as above	2
2. Smooth, brown, constricted at points along its length	*Scytosiphon lomentaria* (plate 15)
2. Rough texture, not constricted	*Asperococcus echinatus* (plate 16)

Enteromorpha intestinalis

Enteromorpha intestinalis	**én - te - ro - mór - fa in - tés - ti - nál - iss**
enteromorpha: gut-shaped	**intestinalis: pertaining to the intestine**

Common Names: "water gut," "sea grass," "maiden-hair," "link confetti"

As the species name implies, *Enteromorpha intestinalis* is a hollow tubular unbranched plant. The bright green tubes often reach several feet in length and in the lower intertidal and shallow sublittoral the actively photosynthesizing plants often entrap oxygen bubbles which account for the periodic swellings seen in the illustration.

Like other *Enteromorpha* species, *Enteromorpha intestinalis* may flourish in polluted areas and can tolerate wide fluctuations in salinity (p. 22).

Similar Species: *Enteromorpha linza* (plate 2) is not tubular throughout.

Range: Florida to Ellesmere Island.

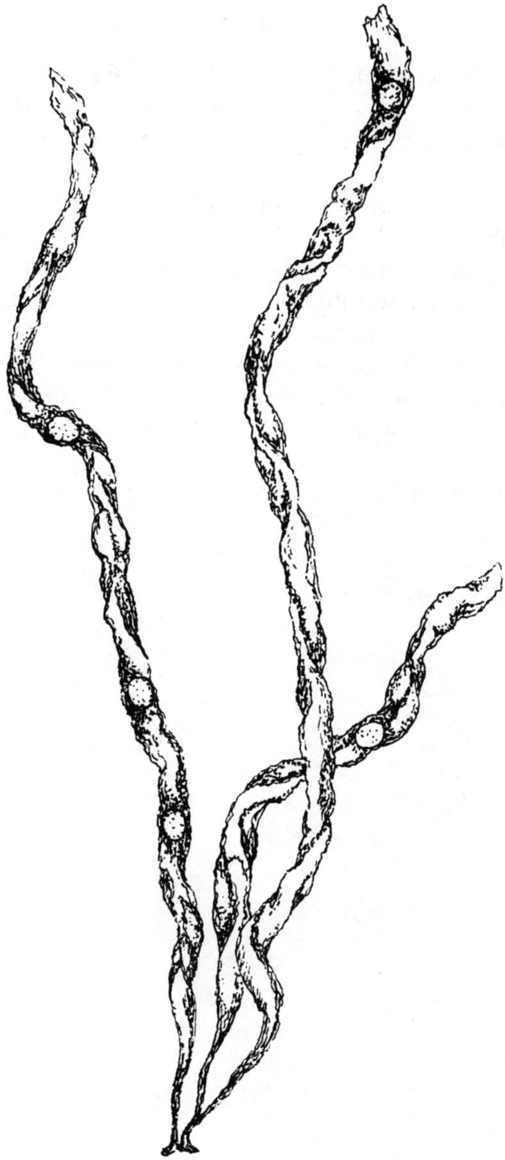

Plate 14. Enteromorpha intestinalis

Scytosiphon lomentaria

Scytosiphon lomentaria	**síght - o - sígh - fon** **ló - men - táy - ree - a**
scytosiphon: leather tube	**lomentaria: mealy**

Scytosiphon, which occurs world-wide in temperate regions, is a leathery unbranched brown tube, 6-18 inches tall, about the thickness of a drinking straw, which tends to grow in clusters on rocks in the lower intertidal and shallow sublittoral down to at least 10 feet. Mature plants are somewhat constricted at intervals along the tubular body, so that the plant may sometimes resemble an irregular chain of narrow sausages. Underwater it can be seen that the tubes are covered with a dense coat of fine colorless hairs.

Similar Species: *Chorda filum* (plate 37) is solid and not constricted.

Range: South Carolina to Northern Labrador.

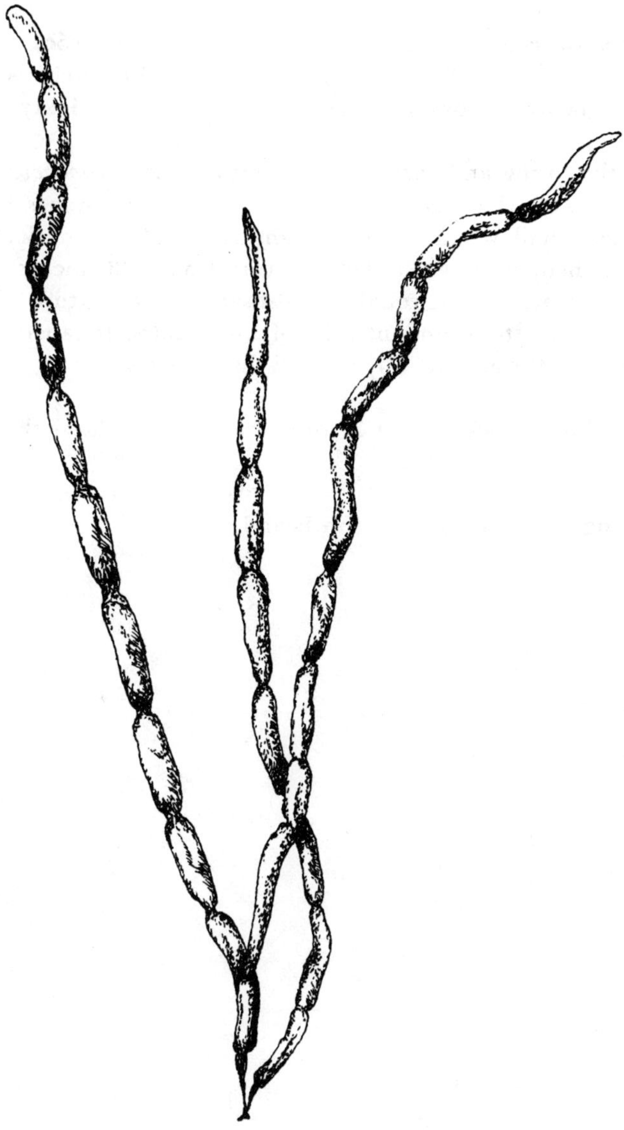

Plate 15. Scytosiphon lomentaria

Asperococcus echinatus

Asperococcus echinatus	**ás - pe - ro - cóck - us** **éck - in - áte - us**
asperococcus: rough-seeded	**echinatus: prickly**

In the spring and early summer, look for *Asperococcus* growing in clusters on the blades of *Fucus* species, or on stones in relatively quiet, shallow water. The plant is a light to dark brown unbranched, narrow tubular structure from 4-20 inches long which has a characteristic rough, almost sandpapery texture. This coarseness is due to abundant tufts of small hairs. It tapers toward the base, and is attached by a small disc holdfast.

Similar Species: Young *Scytosiphon* (plate 15) lacks the rough texture.

Range: New York to Baffin Island.

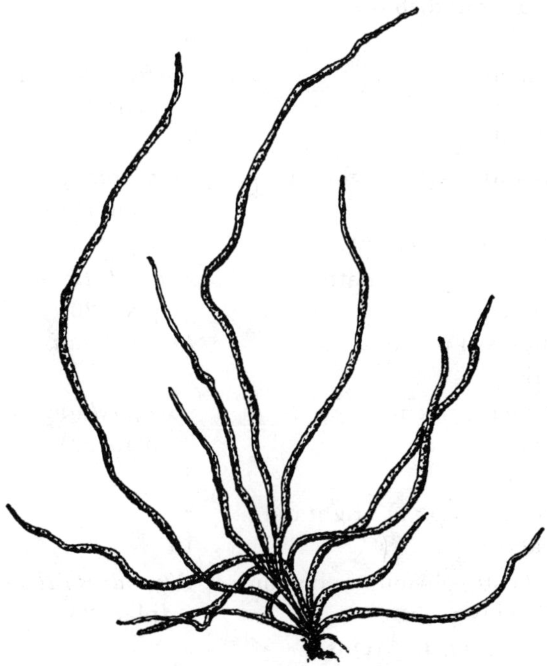

Plate 16. Asperococcus echinatus

KEY TO FLATTENED, BRANCHED SEAWEEDS

1.	Very hard, pink-white segments, branched	*Corallina officinalis* (plate 17)
1.	Not as above	2
2.	Conspicuous midrib present	3
2.	No midrib	6
3.	Air bladders present	*Fucus vesiculosus* (plate 18)
3.	No air bladders	4
4.	Small plants, dark to light red	*Membranoptera alata* (plate 19)
4.	Larger, brownish plants	5
5.	Swollen tips of plants flattened, elongated	*Fucus distichus* (plate 20)
5.	Swollen tips of plants almost spherical	*Fucus spiralis* (plate 21)
6.	Air bladders present	*Ascophyllum nodosum* (plate 22)
6.	No air bladders	7
7.	At least part of plant bright red	8
7.	Not as above	13
8.	Very delicate, obvious "veins" on leaflike branches	*Phycodrys rubens* (plate 23)
8.	Fronds tougher, no "veins"	9
9.	Large fronds, segmented like a palm and fingers	*Palmaria palmata* (plate 24)
9.	Not as above	10
10.	Small narrow flat blades with tiny branched projections	*Rhodophyllis dichotoma* (plate 25)
10.	Not as above	11
11.	Large (to 1 ft.), flattened axes, pointed unbranched projections	*Gracilaria tikvahiae* (plate 26)
11.	Not as above	12

12. Very thin narrow dark branches, expanding into small thin blades	*Phyllophora truncata* (plate 27)
12. Wider, bright red throughout, firm	*Callophyllis cristata* (plate 28)
13. Featherlike branching	14
13. Not as above	15
14. Very fine, abundant branches	*Plumaria elegans* (plate 29)
14. Coarser, one member of branch pair often smaller than opposite member	*Ptilota serrata* (plate 30)
15. Elongated, flattened, tubular in part	16
15. Not as above	17
16. Frequent tiny projections along margins	*Devaleraea ramentaceum* (plate 31)
16. No such projections	*Dumontia contorta* (plate 32)
17. Small, smooth purple-black to greenish blades	*Chondrus crispus* (plate 33)
17. Small dark blades, with conspicuous "knobby" surface toward tips	*Gigartina stellata* (plate 34)

Corallina officinalis

Corallina officinalis **có - ra - lié - na o - fiśs - i - nál - iss**
corallina: a little coral **officinalis: pertaining to drugs**
Common Name: "coral weed"

No other seaweed in our area resembles *Corallina.* The branching filaments making up the body of the plant are coated with calcium and magnesium carbonate which give the plant a very hard outer crust. This is divided into flattened segments, and the whole plant is branched in one plane to give it a roughly pyramidal featherlike shape. Many of these branching axes arise from a single disclike holdfast, and in the shallow sublittoral *Corallina* may cover wide areas of the rocky bottom. The healthy growing plants are purple-red in color, and those which are detached soon fade to a chalky white.

To keep *Corallina,* do not press it, simply air dry it.

Similar Species: none.

Range: Virginia to Newfoundland.

Plate 17. Corallina officinalis

Fucus vesiculosus

Fucus vesiculosus **féw - cuss ve - síck - u - lów - sus**
fucus: paint **vesiculosus: full of little bladders**
Common Names: "bladder wrack," "paddy tang," "rockweed"

This *Fucus* species usually forms a band of abundant growth toward the middle of the intertidal zone. It grows up to 3 feet long and has a prominent midrib in its flattened blades, as well as paired air bladders which serve to keep it afloat, and thus more effectively expose its photosynthetic tissues to sunlight. The plants are either male or female, and the football-shaped structures at the tips of the plant are orange if male and olive-green if female. In sheltered areas, the air bladders are numerous and the fronds are flat and broad, whereas more exposed plants tend to be narrower and bear fewer bladders. Occasionally bladderless forms are found (Note: read the description of *Fucus spiralis*, p. 88).

Similar Species: *Fucus spiralis* (plate 21) has winged receptacles. *Fucus distichus* (plate 20) has long, flattened receptacles.

Range: North Carolina to Labrador.

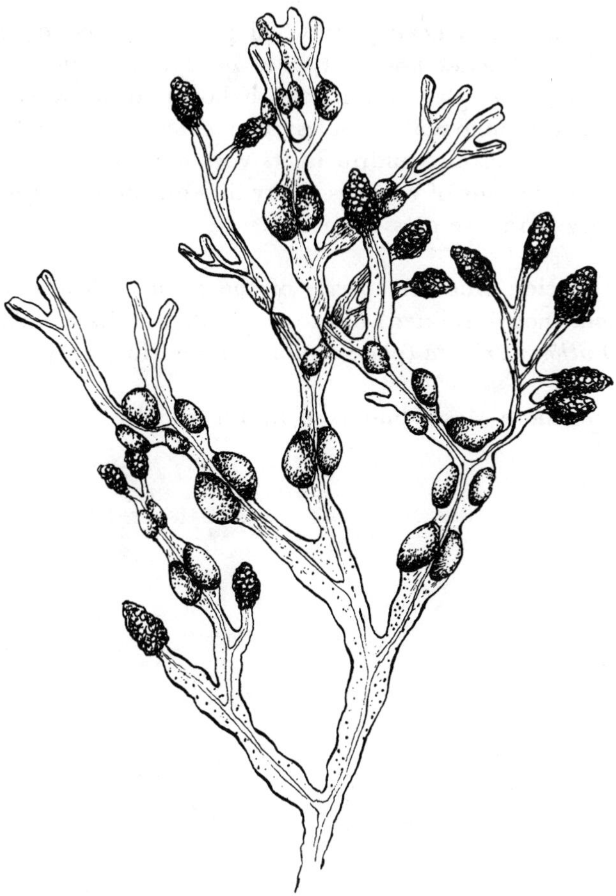

Plate 18. Fucus vesiculosus

Membranoptera alata

Membranoptera alata — **mém - bran - óp - te - ra a - láh - ta**

membranoptera: a membrane — **alata: winged**

This is a beautifully delicate plant which grows attached to stones or larger algae such as *Laminaria* (plates 7, 11-13), *Chondrus* (plate 33) or *Phyllophora* (plate 27) in cold shallow to deep water. It is a small deep crimson to pink plant less than 5 inches high, and unique in that it has a distinct midrib in its very narrow, branched flat blades, on either side of which is a very thin membrane. It makes a particularly attractive pressed specimen.

Similar Species: Another species of the genus, *Membranoptera denticulata* (not illustrated) has toothed edges in its membranous blades. *Euthora cristata* (plate 28) lacks a midrib.

Range: Northern Massachusetts to Baffin Island.

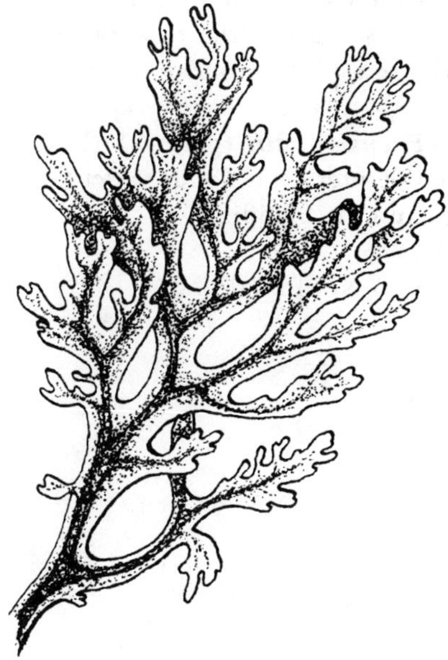

Plate 19. Membranoptera alata

Fucus distichus

Fucus distichus **féw - cuss dís - ti - kus**
fucus: paint **distichus: 2 rows**
Common Name: "sea-wrack"

The habitat and the shape of the receptacles distinguish this from the other *Fucus* species (Note: read the description of *Fucus spiralis,* p. 88). *Fucus distichus* grows on the rocks of the low intertidal and shallow sublittoral and the receptacles are long and flattened. These receptacles are 2-4 inches long and each has a pointed tip. Unlike those of the other *Fucus* species the receptacles are not sharply demarcated from the rest of the fronds on which they are found.

Similar Species: *Fucus spiralis* (plate 21) has winged receptacles and grows high in the intertidal. *Fucus vesiculosus* (plate 18) has inflated receptacles and air bladders.

Range: New Jersey to Newfoundland.

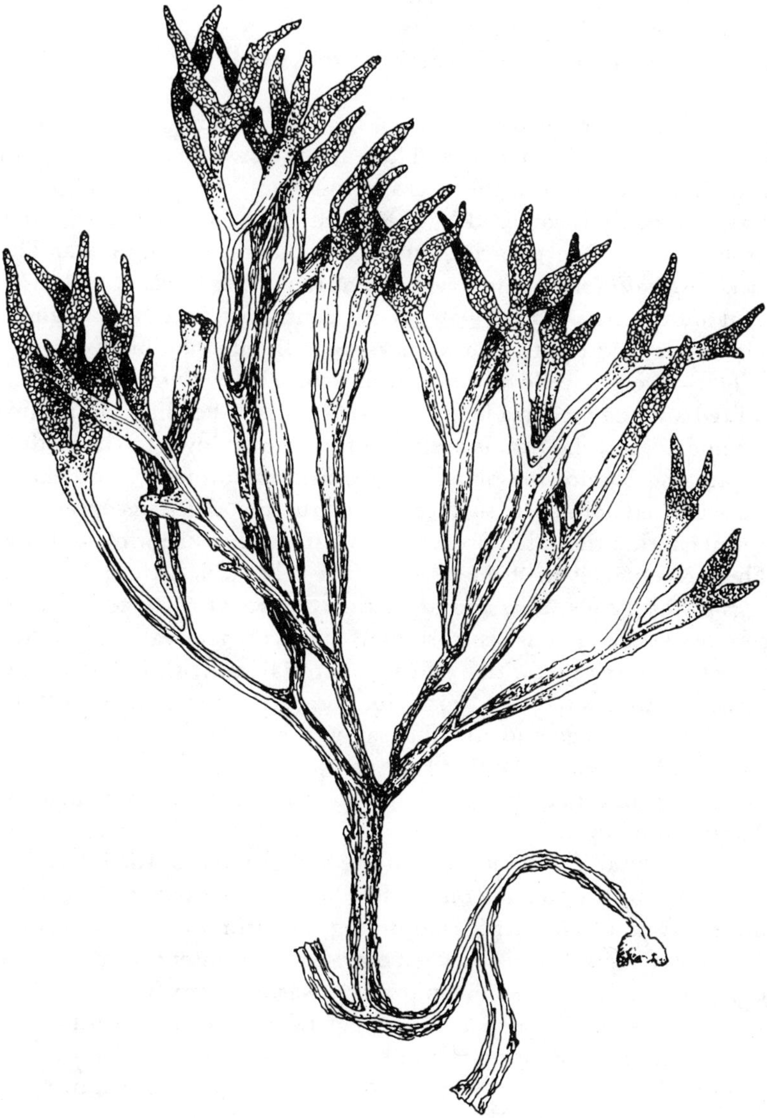

Plate 20. Fucus distichus

Fucus spiralis

Fucus spiralis **féw - cuss spy - rál - iss**
fucus: paint **spiralis: twisted**
Common Names: "spiral wrack," "rockweed"

Even the casual observer of our seashore life could not fail to notice at lease one species of *Fucus.* These large brown seaweeds grow in abundance and cover the rocks throughout the intertidal zone. They are easily recognized by their numerous saclike reproductive structures called receptacles, growing at the tips of the plants. The species *spiralis* forms a narrow band at the upper levels of the intertidal and like the other *Fucus* species consists of flattened, branching fronds held tightly to the rocks by a disclike holdfast.

The olive- to yellow-brown fronds of *Fucus spiralis* are often twisted and may reach a length of 6-18 inches. They have a prominent midrib and the fronds are often worn away down to the midrib toward the base of the plant. Numerous tufts of dark brown hairs are scattered across the surface. The structure of the receptacles is unique in that they are bordered by a narrow shelflike wing of tissue. These receptacles hold the reproductive cells, which are sperm and eggs in this genus, and small spherical masses of these cells can be clearly seen and felt as protrusions in the receptacle wall. Inside, the receptacles are filled with a jellylike material. In Irish folk medicine this came to be used as a "cure" for corns. The receptacles or "jelly bags" were crushed and added to salt water used to soak the afflicted feet. In 18th century England the receptacles were added to grapes and sugar and used in fermenting wine to be used "to treat bone diseases in children."

Fucus spiralis, being located in the higher levels of the intertidal, is covered with water for only a few hours in the day and hence is quite resistant to damage from drying out. In reproduction, as the *Fucus* plants slowly dry out when exposed, the contents of the receptacles ooze out, and the incoming tide washes them from the plant surface. Fusion of the sperm and eggs takes place in the water. As with most algae, enormous numbers of reproductive cells must be produced because the chances of fertilization and survival are very slim in an aquatic environment.

Species of *Fucus* do not adhere to paper and must first be dried and then fastened by tape or glue (p. 32).

Similar Species: *Fucus vesiculosus* (plate 18) has air bladders and lacks the winged receptacles. *Fucus distichus* (plate 20) has long, flat receptacles.

Range: New York to Newfoundland.

Plate 21. Fucus spiralis

Ascophyllum nodosum

Ascophyllum nodosum **ásk - o - fíll - um no - dóse - um**
ascophyllum: sac leaf **nodosum: full of swellings**
Common Names: "knotted wrack," "yellow tang," "sea whistle," "knot weed"

This is one of the large seaweeds which are so characteristic of the rocky coasts of our colder waters. *Ascophyllum* is a plant of the intertidal zone, and where the water is relatively calm it may grow abundantly. It is not easily confused with any other species because of its narrow, straplike olive-green blades which bear scattered air bladders which grow singly and may reach the size of a walnut, although they are usually smaller. It is said that after one or two years *Ascophyllum* forms one air bladder for each year's growth. It is reported to have a maximum life span of fifteen years. A single plant, attached by a disclike holdfast, branches irregularly and is usually 2-3 feet long.

During the winter the reproductive process occurs, and small pea-sized yellow structures called receptacles may be found along the length of the plant, attached by short stalks. After the sperm and eggs are released from the cavities of the receptacles the latter drop off leaving the supporting stalks which persist throughout the year. *Ascophyllum* is one of the seaweeds which have both male and female plants, and on the female the receptacles are olive-yellow.

Ascophyllum prefers shores protected from heavy wave action and is absent or reduced under such turbulent conditions. It may also be found attached to rocks in tidal portions of salt marshes. Since it is so large and tends to lie flat at low tide, *Ascophyllum* may cover other intertidal seaweeds. Careful probing around the bases of the plants will reveal these other organisms. There is commonly a small red alga, *Polysiphonia lanosa* (plate 62), growing as small wiry tufts on the fronds of this species.

Similar Species: *Fucus vesiculosus* (plate 18) has a distinct midrib and the receptacles are at the ends of the fronds.

Range: Virginia to the Arctic.

Plate 22. Ascophyllum nodosum

Phycodrys rubens

Phycodrys rubens **fy - có - driss réw - bens**
phycodrys: seaweed **rubens: reddish**
Common Name: "seaweed oak"

Few seaweeds are better for the preparation of attractive pressed specimens than *Phycodrys*. It is very common, especially north of Cape Cod, and its delicate pink-red blades are unmistakable because they have a conspicuous midrib and distinct "veins" which resemble the vein systems in the leaves of land plants such as the oak or maple. These are, of course, not really veins, which act as functional conducting tissue, but merely "decorative." Apart from the midrib and veins the blades, usually about 1/2 inch wide, are only 1 cell in thickness. *Phycodrys* grows to depths of at least 75 feet on stones and shells and is frequently cast up on shore. Plants collected in late summer and autumn may be covered with a gray feltlike tissue—this is a colony of bryozoans, which are invertebrate animals that commonly attach to seaweeds.

Similar Species: none.

Range: New Jersey to the Arctic.

Plate 23. Phycodrys rubens

Palmaria palmata

Palmaria palmata — **pahl - máre - ee - a pahl - máh - ta**

palmaria: like a palm — **palmata: like the palm**

Common Name: "dulse"

This attractive edible seaweed is easily recognized. It is a flat, leathery purple-red frond which may branch several times and often has small "proliferations" or outgrowths along the edge, particularly toward the base where there is a small disc holdfast. The plant may reach 20 inches in length and is found year-round growing attached to or near *Laminaria* (plates 7, 11-13). As described in the Introduction (p. 34) it is collected in Maine and the Maritime Provinces, dried and eaten as a kind of "seaweed potato chip." In addition to being considered by some as a delicacy, it makes an admirable pressed specimen.

Similar Species: none.

Range: New Jersey to the Arctic.

Plate 24. Palmaria palmata

Rhodophyllis dichotoma

Rhodophyllis dichotoma **ró - doe - fíll - iss die - cót - o - mah**
rhodophyllis: red leafed **dichotoma: split in two**

This is a beautiful little plant that should be immediately recognizable because the margins of the small flat blades are lined with tiny branched toothlike projections. The rosy red or red-purple blades are only 2-5 inches long and are narrow and delicate. Look for this in deep water attached to shells, stones and occasionally to other seaweeds. It retains its color very well when pressed.

Similar Species: *Euthora cristata* (plate 28) has more highly branched blades and lacks the marginal projections.

Range: Northern Massachusetts to Ellesmere Island.

Plate 25. Rhodophyllis dichotoma

Gracilaria tikvahiae

Gracilaria tikvahiae **gráss - i - lá - ree - a tík - vah - yay**
gracilaria: slender **tikvahiae: after Tikvah Edelstein, a Canadian botanist**

The irregularly branched axes of *Gracilaria tikvahiae* are of a faded purple hue, and are flattened to a width of usually under 1/2 inch. The margins bear numerous short pointed branches, and all of the branching is in one plane so that the entire plant is relatively flat and only slightly bushy. The plant grows in shallow water just below the low tide line. It does not adhere well to paper and should be taped or glued after drying. World-wide there are at least 100 species of *Gracilaria*. The species *confervoides* is abundant in Australia and is collected there for the extraction of agar (p. 40).

Similar Species: *Agardhiella* (plate 54) is cylindrical and the branches are tapered at both ends.

Range: Florida to Eastern Canada.

Plate 26. Gracilaria tikvahiae

Phyllophora truncata

Phyllophora truncata **fill - áh - fo - ra trun - káy - ta**
phyllophora: leaf-bearing **truncata: cut off**

There are 4 species of this genus reported from our area. *Phyllophora truncata*, illustrated here, and *Phyllophora pseudoceranoides* are the most common. The former is a small plant, up to 4 inches tall, consisting of one to several thin, occasionally flattened, branched stalks which expand into narrow wedge-shaped membranous blades. These blades are usually forked several times and are an attractive red-purple. They flourish in the spring attached to other seaweeds, and on rocks or shells, to at least 70 feet in depth.

Phyllophora pseudoceranoides may be somewhat large and its stalk and lower branches are characteristically cylindrical. The dull purple blades are firm and bear several sharp points. It too is a plant of deep water but may also be found in tide pools.

Similar Species: *Chondrus crispus* (plate 33) may resemble *Phyllophora pseudoceranoides*, but the entire frond is flattened, in contrast to the cylindrical stalks of the latter.

Range: New Jersey to Newfoundland.

Plate 27. Phyllophora truncata

Callophyllis cristata

Callophyllis cristata **cal - low - fíll - iss** **kris - táy - ta**
callophyllis: beautiful leaf **cristata: crested**

Small though it may be, this 1-2 inch plant is one of the more beautiful of the seaweeds, and makes an excellent specimen for pressing. It is bushy in habit, but the individual fronds are flattened and repeatedly branched, giving a rather intricate webbed appearance. The color is a striking rose-pink to dark red, and it can be found by carefully examining the bases of *Laminaria* (plates 7, 11-13) or other species such as *Phyllophora* (plate 27) to which it attaches in clusters. It is typically a plant of deeper water but may be found in tide pools.

Similar Species: *Ptilota serrata* (plate 30) also attaches to the base of *Laminaria,* but is a larger dark brown plant which is more abundantly branched, and often one member of the branch pair is much smaller than the other.

Range: New Jersey to the Arctic.

Plate 28. Callophyllis cristata

Plumaria elegans

Plumaria elegans	**plu - máy - ree - a él - egg - ans**
plumaria: feathery	**elegans: elegant, fine**

Like *Ptilota* (plate 30) and *Euthora* (plate 28) this seaweed prefers to grow attached to the bases of larger plants such as *Laminaria* (plates 7, 11-13) or *Phyllophora* (plate 27). It too consists of flat branching fronds, but these are much more abundantly branched and finer than the above, although the dark brown to red-purple color is quite similar to that of *Ptilota*. Toward the northern part of its range it becomes intertidal, growing under overhanging *Fucus* (plates 18, 20, 21), and always stays away from direct sunlight, preferring shaded perpendicular or overhanging rocks.

Similar Species: *Ptilota serrata* (plate 30) is more coarsely branched.

Range: New Jersey to Newfoundland.

Plate 29. Plumaria elegans

Ptilota serrata

Ptilota serrata **tie - lów - ta ser - ráh - ta**
Ptilota: plumed **serrata: saw-toothed**
Common Name: "feather weed"

This is another one of those smaller seaweeds that seem to prefer to grow attached about the base of *Laminaria* (plates 7, 11-13), so that one must hunt for it by cautiously separating the large blades and searching among the stipes and holdfasts for specimens. *Ptilota* makes an excellent pressed specimen because it grows as flat feathery fronds. These fronds are dark brown to red-purple and grow in bushy tufts up to 6 inches long. The small flat branches are generally opposite one another, and often the growth of one member of an opposite pair is suppressed so that it is much smaller than its partner.

Similar Species: *Euthora* (plate 28) is rose-pink to dark red and has a different branching pattern. *Plumaria* (plate 29) has more abundant, finer branches.

Range: New Jersey to Ellesmere Island.

Plate 30. Ptilota serrata

Devaleraea ramentaceum

Devaleraea ramentaceum	**de - va - lé - ree - a ra - men - táy - see - um**
devaleraea: after M. DeValera, an Irish botanist	**ramentaceum: pertaining to branching**

Some seaweeds can be perplexingly variable in their appearance, which may fluctuate depending upon the environmental conditions under which they are growing. In relatively quiet water, *Devaleraea* grows as a cluster of tubular, flattened fronds attached by a small holdfast. There may be a few or many simple branches near the base. In cold water where there are swift currents, such as in northern Maine, these fronds "proliferate," that is, they develop numerous short branches all along their length. In either case the plant is a dark brown to red-purple and may reach a length of 16 inches, though it is usually much shorter. It develops on stones and shells in shallow water and is of a firm, coarse consistency.

Similar Species: *Dumontia contorta* (plate 32) may resemble the sparsely branched forms, but its branches are longer and often quite twisted, and the branch ends are often frayed. Also, *Dumontia* is found as far south as Long Island.

Range: Northern Massachusetts to Ellesmere Island.

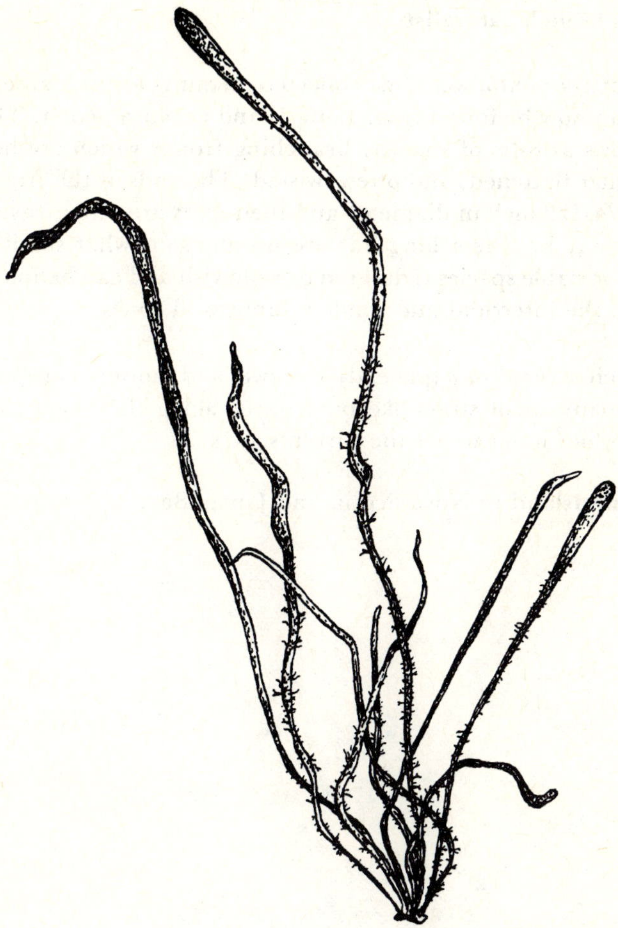

Plate 31. Devaleraea ramentaceum

Dumontia contorta

Dumontia contorta	**dew - món - tee - a** **con - tór - ta**
dumontia: after M. Dumont, a French naturalist	**contorta: twisted**

Specimens of *Dumontia* were first collected on our coast in Maine in 1913 and can now be found from Long Island to Nova Scotia. This seaweed grows as tufts of sparsely branching fronds which are hollow, somewhat flattened, and often twisted. The ends of the fronds are about 1/4-1/2 inch in diameter and their ends are often frayed. The plants may be 2 feet long but are usually somewhat smaller. This rather variable species is dull red or yellowish and can be found on stones in the intertidal and shallow sublittoral zones.

Similar Species: *Devaleraea* (plate 31) has fewer and shorter blades and may have many small strandlike outgrowths along the fronds, especially in colder water where the currents are swift.

Range: Long Island to Nova Scotia and James Bay.

Plate 32. Dumontia contorta

Chondrus crispus

Chondrus crispus **kón - drus kris´ - pus**
chondrus: cartilage **crispus: curled**
Common Names: "Irish moss," "carragheen"

The many uses of this economically important seaweed are discussed in the Introduction (p. 35). It grows abundantly on rocks from just below the tide mark down to at least 70 feet. A disclike holdfast supports several flattened branched fronds which are usually 3-4 inches long, and the plants ordinarily grow in large colonies which may cover thousands of square feet of ocean bottom. In deep water the fronds are quite long and dark purple-red, while in shallow water or in tide pools they are short, compact and of a pale green-red hue. Sunlight striking the fronds often gives them a curious irridescent metallic sheen.

Please note on page 37 directions for the preparation of a simple pudding from *Chondrus*.

Similar Species: *Gigartina stellata* (plate 34) usually has a "bumpy" or rough surface toward the ends of the fronds.

Range: New Jersey to Newfoundland.

Plate 33. Chondrus crispus

Gigartina stellata

Gigartina stellata — **ji - gár - ti - na stel - láh - ta**
gigartina: a grape stone — **stellata: starry**

This seaweed grows in the fringe of the low intertidal zone and closely resembles *Chondrus crispus* (plate 33) with which it is often found. It ordinarily grows in thick tufts 2-3 inches tall, and its disc-like holdfast supports a twisted, somewhat flattened axis. The terminal surfaces of the dark purple-red flattened blades are covered with irregular "lumps" which enclose microscopic reproductive cells. *Gigartina stellata* is our only species, but the genus *Gigartina* includes over 90 species world-wide and includes the largest of the red algae, some of which may be 6 feet long and 3 feet wide. During World War II *Gigartina* was used in England as a source of agar (p. 40).

Similar Species: *Chondrus crispus* (plate 33) is larger and lacks the irregular outgrowths on the blade.

Range: Rhode Island to Newfoundland.

Plate 34. Gigartina stellata

KEY TO UNBRANCHED, CORDLIKE SEAWEEDS

1. Light to dark green	2
1. Brown	3
2. Bright green, like tangled nylon line	*Chaetomorpha linum* (plate 35)
2. Dark green, grows erect, attached. Segments barely visible.	*Chaetomorpha mela-gonium* (plate 36)
3. Smooth cord, may have short hairs	*Chorda filum* (plate 37)
3. Densely covered with long hairs	*Chorda tomentosa* (plate 38)

Chaetomorpha linum

Chaetomorpha linum **kéy - ta - móre - fa lié - num**
chaetomorpha: bristle-shaped **linum: a net**

There should be no mistaking this seaweed for any other. It looks exactly like a tangled piece of bright green nylon fishing line. It can often be found floating or entangled down to at least 15 feet, and in tide pools. It is actually a long, unbranched chain of cells attached end to end, although the individual cells cannot be seen without some magnification.

Similar Species: *Rhizoclonium* (plate 60) is also green and tangled but is much thinner. *Cladophora* (plate 58) is also thinner and abundantly branched. *Spongomorpha* (plate 59) is also abundantly branched.

Range: Florida to Newfoundland.

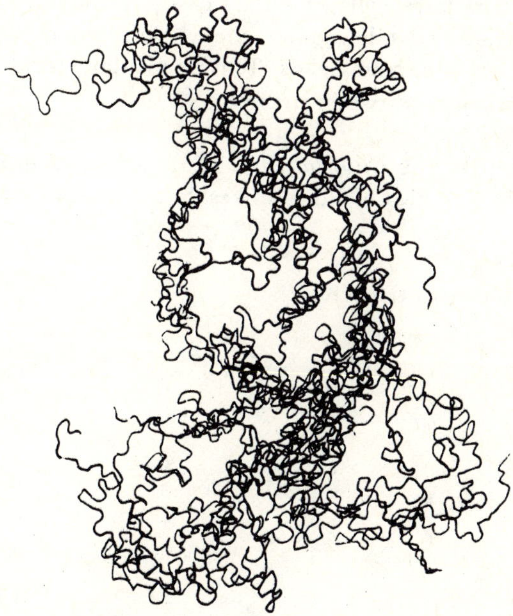

Plate 35. Chaetomorpha linum

Chaetomorpha melagonium

Chaetomorpha melagonium	**kéy - ta - móre - fa** **mél - a - gó - nee - um**
chaetomorpha: bristle-shaped	**melagonium: angled cheeks**

Looking like a dark green spike, this plant grows in scattered groups, often on sand-covered rocks at depths of 3 feet or more. It is unusual because it is an unbranched filament made up of a uniseriate row of very large cells which are easily visible to the unaided eye. They are slightly swollen so that the whole filament looks somewhat like a chain of narrow beads. This particular species will usually not adhere to paper when pressed and dried.

Similar Species: Young *Chorda filum* (plate 37) or *Chorda tomentosa* (plate 38) are also slender spikes but are dark brown and smooth.

Range: New Jersey to the Arctic.

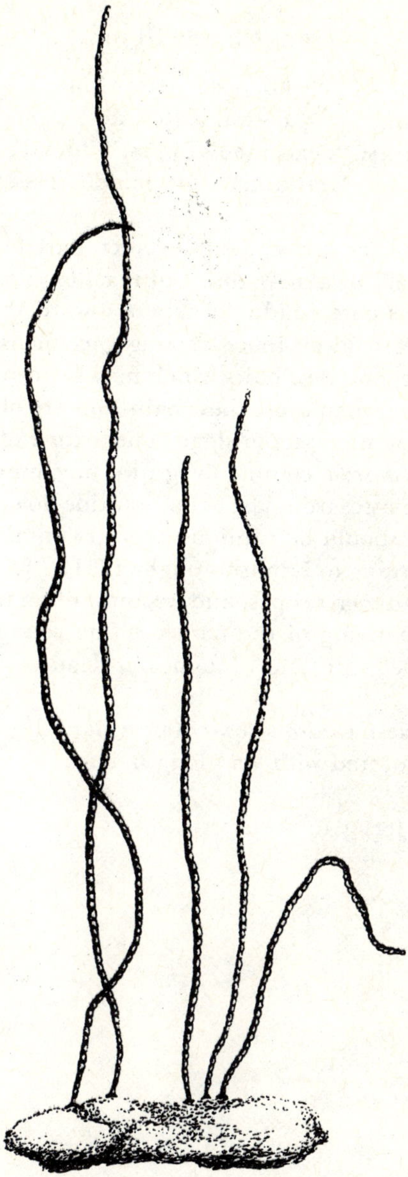

Plate 36. Chaetomorpha melagonium

Chorda filum

Chorda filum **kór - da fié - lum**
chorda:a string **filum: a thread**
Common Names: "dead men's ropes," "devil's shoelace," "sea lace," "mermaid's tresses"

This plant grows as a firm, flexible, erect, dark brown cord attached to stones or shells by a small disc and reaching up to 15 feet or more in length. This cord gradually tapers toward the tip and is often frayed and worn away there. Young specimens are clothed with short, delicate, colorless hairs which may be seen more easily under the water. Older plants lose these hairs and are often covered with a "dusty" layer of microscopic algae. The texture of the plant is somewhat slimy. *Chorda* commonly grows in clumps and may form rather dense masses from just below low tide down to at least 30 feet, and the diver should be cautious about swimming through such a growth. According to Landsborough (1851), "In England it is often called . . . dead men's ropes, and we know of an instance in which it proved too deserving of the name: a fine young man in bathing, being entangled in it and brought out dead."

Similar Species: *Chorda tomentosa* (plate 38) is a spring plant, shorter and covered with very long hair.

Range: New Jersey to the Arctic.

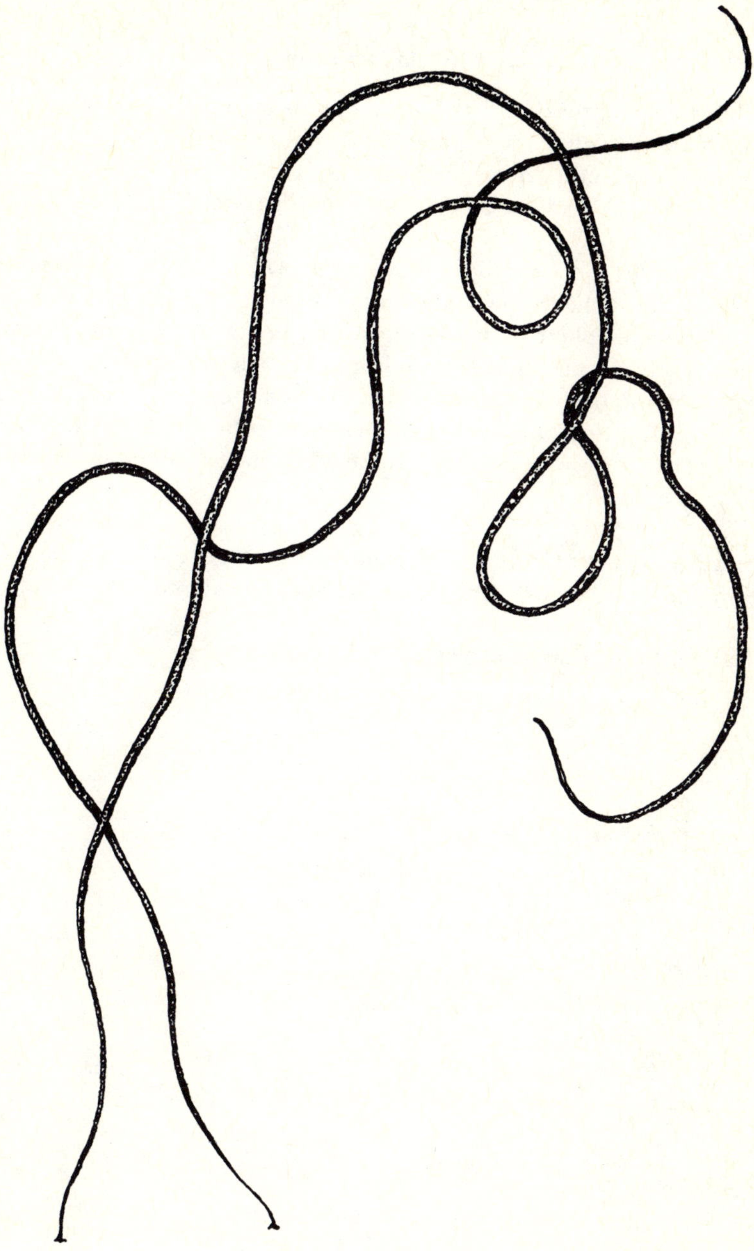

Plate 37. Chorda filum

Chorda tomentosa

Chorda tomentosa **kór - da toe - men - tóe - sa**
chorda: a string **tomentosa: dense hair**
Common Names: "dead men's ropes," "devil's shoelace," "sea lace," "mermaid's tresses"

Like its sister species *Chorda filum* (plate 37), this seaweed is a solid, erect, unbranched brown cord, attached by a small disc. However, it is a smaller plant than *Chorda filum,* being usually about 3 feet long when mature, and is found only in early spring. It is covered with a dense layer of fine hairs up to 1 inch long. Like the shorter hairs of *Chorda filum,* these can be seen best by the diver since they collapse and become scarcely visible when the specimen is removed from the water.

Similar Species: *Chorda filum* (plate 37) is longer and lacks the covering of long hairs. It grows throughout the summer.

Range: Long Island to Labrador.

Plate 38. Chorda tomentosa

KEY TO BRANCHED, CORDLIKE SEAWEEDS

1.	Ropelike forking green strands	*Codium fragile* (plate 39)
1.	Not as above	2
2.	Plant having stiff, wiry texture	3
2.	Plant softer	7
3.	Very dense clusters of short, dark branches on the axes	*Cladostephus spongiosus* (plate 40)
3.	Not as above	4
4.	Large bushy plant, numerous alternate small spiny projections	*Desmarestia aculeata* (plate 41)
4.	Not as above	5
5.	Very rough surface, plant brittle	*Stilophora rhizodes* (plate 42)
5.	Not as above	6
6.	Sturdy, very regularly forked black wiry axes	*Polyides rotundus* (plate 43)
6.	Thinner, irregular forking, yellow when dry	*Ahnfeltia plicata* (plate 44)
7.	Fine filamentous branches, hook-like structures present	8
7.	Not as above	9
8.	Hooks on ends of branches	*Hypnea musciformis* (plate 45)
8.	Hooks along length of branches	*Bonnemaisonia hamifera* (plate 46)
9.	Very slimy, wormlike axes branched sparingly	10
9.	Not as above	11
10.	Solid, dark, slender plant	*Nemalion helminthoides* (plate 47)
10.	Pale brown, very soft axes, about 1/4" in diameter	*Eudesme virescens* (plate 48)

11. Bright red, branches covered with very delicate hairs	*Dasya baillouviana* (plate 49)
11. Not as above	12
12. Large light brown bushy plant, numerous opposite branches	*Desmarestia viridis* (plate 50)
12. Not as above	13
13. Most branches from main axis unbranched and longer. Plant dark, slippery texture	*Chordaria flagelliformis* (plate 51)
13. Not as above	14
14. Dull red-brown, gelatinous consistency, of numerous small, cask-shaped segments	*Champia parvula* (plate 52)
14. Not as above	15
15. Main axis much wider than the bushy branches	*Cystoclonium purpureum* (plate 53)
15. Not as above	16
16. Bushy, translucent, up to 1 foot tall, axis and branches cylindrical	*Agardhiella subulata* (plate 54)
16. Not as above	17
17. Branches alternate or on one side, tapering near each end, curved	*Lomentaria baileyana* (plate 55)
17. Branches not curved	*Chondria* sp. (plate 56)

Codium fragile

Codium fragile | **kó - dee - um frá - jee - lay**
codium: animal skin | **fragile: delicate**
Common Names: "sea staghorn," "oyster thief"

The story of *Codium*'s sudden appearance in Long Island Sound and its rapid spread to southern Cape Cod is a fascinating one and as yet not totally explained. It was first seen in 1957 at Montauk Point at the eastward end of Long Island. Since that time this "weedy" species has spread and is now a dominant member of the sublittoral vegetation. *Codium fragile* probably arrived on a shipment of oysters from Europe, where it has also became a "weed." In 1962 it arrived at Cape Cod from Long Island Sound in the same manner. It can be found at depths from 1 to at least 40 feet, depending upon light penetration, and its form is unmistakable. Developing first as a spongy green crust, it soon elongates into long, erect branching cylindrical fronds looking very much like green ropes. Large plants are about 3 feet long and the entire plant may be quite massive. The genus *Codium* is a common one in warm temperate and subtropical waters and a Mexican species, *Codium magnum,* may reach a length of 26 feet.

The construction of the plant body of *Codium* is quite unusual for seaweeds of our waters since the entire plant is a series of densely interwoven, interconnected filaments which have no internal cross walls and hence are really very large single cells. The spongy rope-like strands have a colorless interior and an exterior which is dark green because the photosynthetic pigments concentrate toward the ends of the filamentous cells. The growing tips are a lime green and the entire plant when seen underwater is covered by a "fuzzy" white outer layer of very fine hairs.

The study of *Codium*'s growth and distribution is of more than academic interest since this plant, which needs a solid substrate on which to attach and grow, often attaches to scallops and oysters on the sea bottom. Oyster beds in shallow bays, ordinarily free from interference from large seaweeds, are often infested with *Codium,* which finds them to be an excellent substrate. *Codium* plants on mature oysters must be removed before they can be shipped, adding labor and time to the process which results in higher prices. Also, when the plants are actively photosynthesizing in shallow water, enough oxygen bubbles may get trapped within the plant to make it buoyant and cause it to rise to the surface, carrying the hapless oyster along with it.

North of Cape Cod, *Codium* has been reported in Boothbay Harbor and at Appledore Island, Maine. Perhaps some observant diver will soon report its occurrence in other areas if it continues to spread as it has in the recent past.

Similar Species: none.

Range: Barnegat Bay, New Jersey, to Southern Cape Cod. Also in Boothbay Harbor and at Appledore Island, Maine.

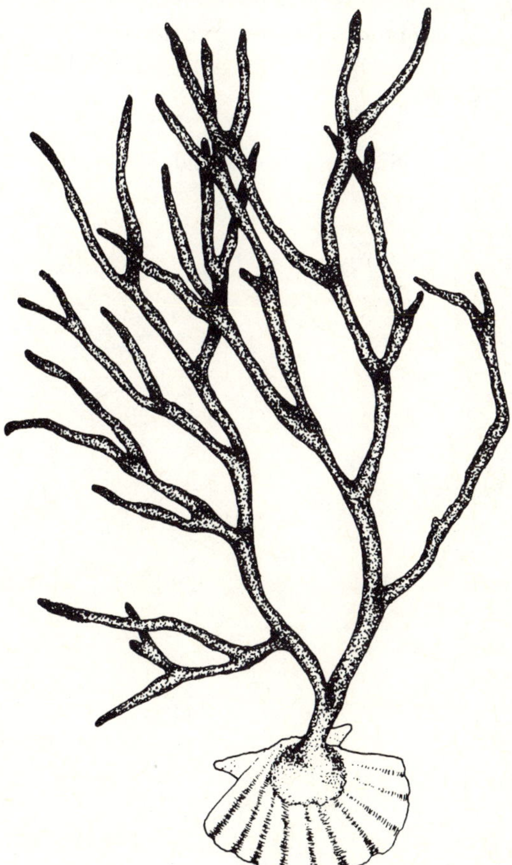

Plate 39. Codium fragile

Cladostephus spongiosus

Cladostephus spongiosus	**klá - doe - stéf - us** **spún - gee - ó - sus**
cladostephus: branch crown	**spongiosus: spongy**

Note the very restricted range of this plant. Perhaps this is one of those seaweeds whose known range will be extended by observant divers. It is usually 4 or 5 inches tall and attached in the sublittoral down to at least 35 feet. It is somewhat stiff, spongy and dark brown with rather sparse, spreading branches which are covered by dense clusters of very fine filaments. These filaments are usually absent toward the base of the plant so its lower portions are bare and wiry.

Similar Species: none.

Range: Long Island to Southern Cape Cod.

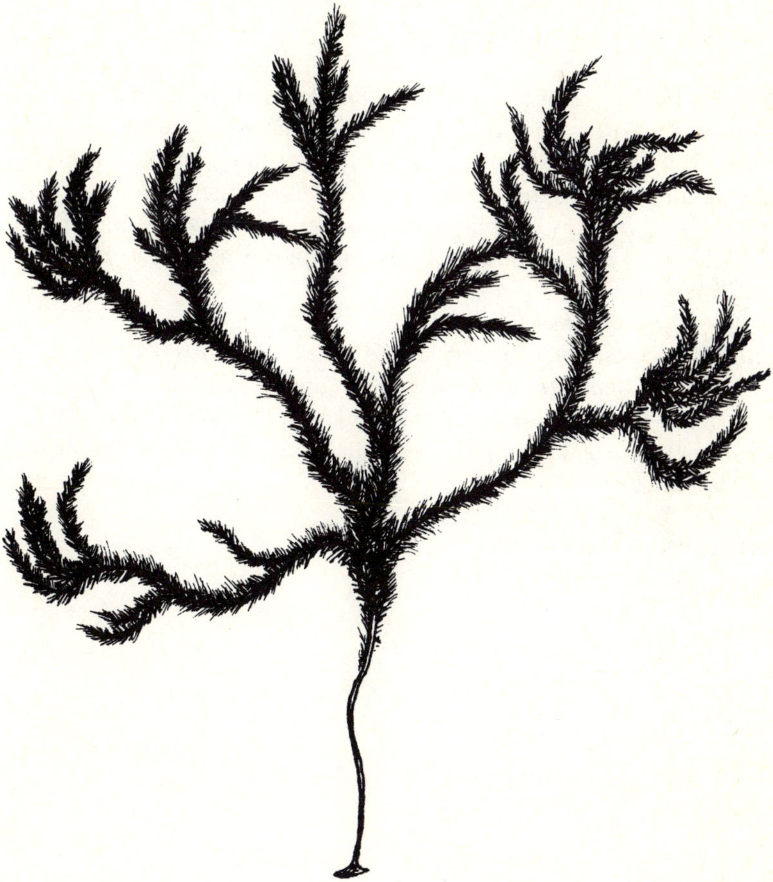

Plate 40. Cladostephus spongiosus

Desmarestia aculeata

Desmarestia aculeata — **déz - ma - rés - ti - a a - kyú - lee - ah - ta**

desmarestia: after Desmarest, a French naturalist — **aculeata: spiny**

Common Name: "landlady's wig"

A finely divided, alternately branching plant, *Desmarestia aculeata* has a stiff texture, and its dark brown flattened branches are covered in the spring with tufts of small brown hairs which fall off by midsummer leaving small projections. It may form extensive beds on stones down to at least 50 feet. It is usually between 1 and 2 feet long although some related species outside our area may reach 14 feet in length.

Similar Species: *Chordaria* (plate 51) has wider, usually undivided branches. *Desmarestia viridis* (plate 50) has opposite branching and the branches are cylindrical. *Dictyosiphon* (plate 67) has much thinner branches. All of the above lack the spiny processes of *Desmarestia aculeata.*

Range: New Jersey to Newfoundland.

Plate 41. Desmarestia aculeata

Stilophora rhizodes

Stilophora rhizodes	**still - áh - for - a ry - zó - deez**
stilophora: dot-bearing	**rhizodes: resembling a root**

Stilophora prefers quiet, relatively warm, protected bays and low water down to perhaps 15 feet. It may reach 1 foot in length in the early spring and summer, and because of its rather sparse irregular branching it does not have an obvious main axis. The plant is pale brown and has a characteristically rough, granular surface due to small clumps of hairs. The branches are somewhat stiff and brittle. *Stilophora* is often found unattached and drifting.

Similar Species: *Ahnfeltia* (plate 44) is stiff and wiry but is not brittle and is much thinner and smoother in texture.

Range: North Carolina to Prince Edward Island.

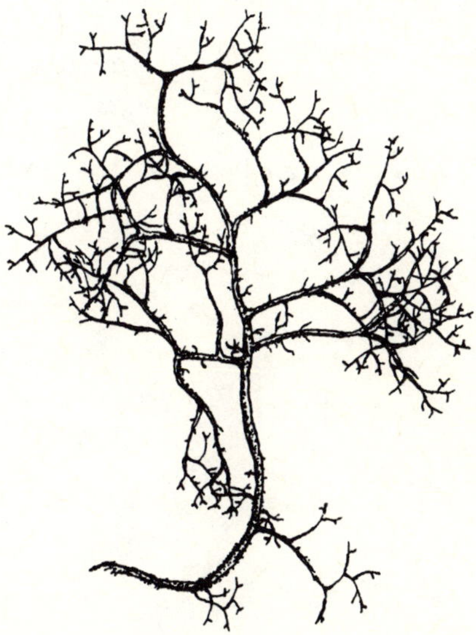

Plate 42. Stilophora rhizodes

Polyides rotundus

Polyides rotundus **póll - lee - éye - deez** **ro - tún - dus**
polyides: many-formed **rotundus: round**
Common Name: "goat tang"

Botanists used to see *Polyides* only when they examined material dredged from the ocean bottom in deep water. Now, with diving gear, one can find this seaweed attached and study it more carefully. The plant grows as several erect, stiff-cylindrical stalks arising from a single disc attached to a rock. These stalks branch 6-8 times in a regular forking pattern and they may reach 1 foot in height. *Polyides* is dull red but when dry turns black.

Similar Species: *Ahnfeltia* (plate 44) is a thinner, more wiry plant which is more abundantly branched and dries to a yellow-white.

Range: Long Island to Labrador.

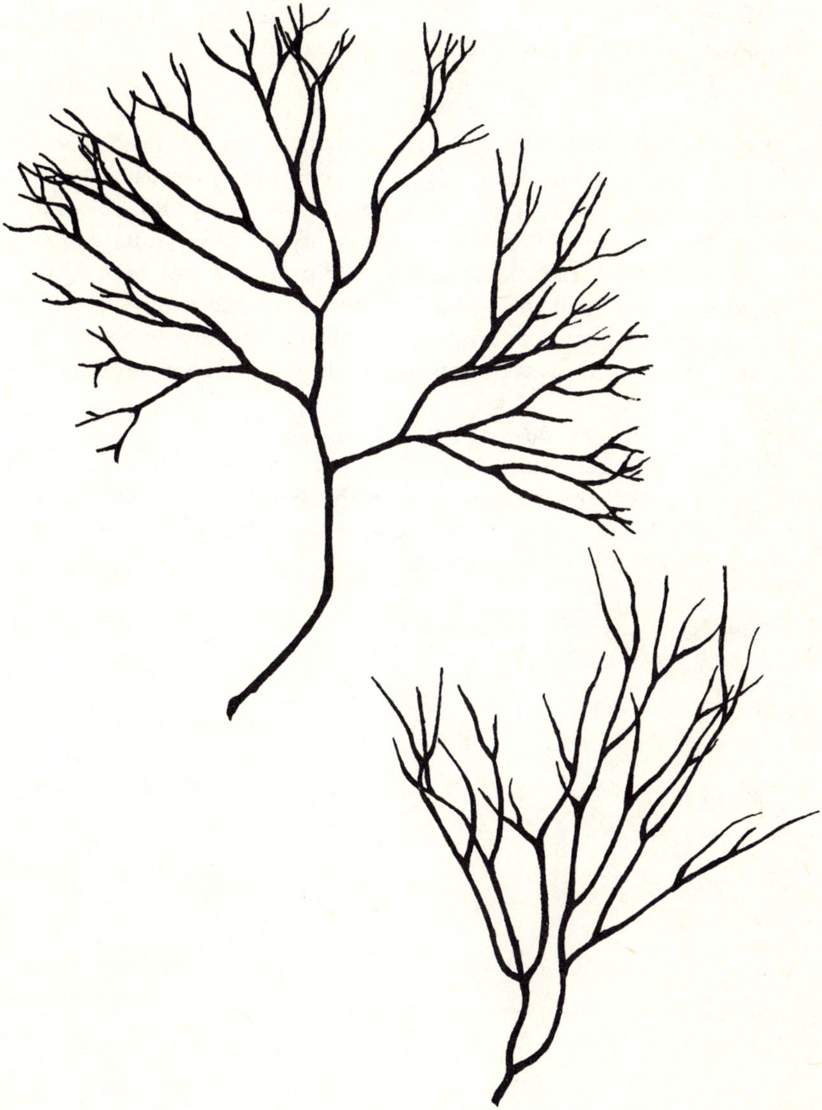

Plate 43. Polyides rotundus

Ahnfeltia plicata

Ahnfeltia plicata — **on - féll - tee - a pli - káy - ta**

ahnfeltia: after N.O. Ahnfelt, a Swedish botanist — **plicata: folded up**

Ahnfeltia is one of those seaweeds which is definitely seen to best advantage when diving. It often grows at depths between 40-50 feet and seems to prefer silty crevices between rocks or cobbles. It has a very stiff, wiry consistency and grows in rigid, dense tufts which branch repeatedly, 10 or more times. The branches are about 1/25 inch in diameter, and the tips are blunt. *Ahnfeltia* is often found tossed up on the beach, and in this condition it is usually bleached yellow-white. It will press flat but will usually not adhere to paper.

Similar Species: *Polyides* (plate 43) is less branched, and the branches are thicker. *Stilophora* (plate 42) is thicker, quite brittle, and has a rough texture. Sometimes the holdfasts of *Laminaria* (plates 7, 11-13) or the roots of *Zostera* (eel-grass) may be found and mistakenly identified as *Ahnfeltia*.

Range: New Jersey to Devon Island.

Plate 44. Ahnfeltia plicata

Hypnea musciformis

Hypnea musciformis	**híp - knee - a mús - i - fór - mis**
hypnea: a mass	**musciformis: mosslike**

This is not a common plant in the northern part of its range but is quite distinctive because of the swollen, hooked ends of its long branches. *Hypnea* grows on rocks in relatively warm protected areas down to at least 20 feet as a fleshy textured plant whose cylindrical main axis bears occasional side branches, which themselves are covered with relatively short, irregularly pointed branches. The color is dull purple or purple-red but is often bleached and strawlike. Mature plants reach 20 inches. Species of the genus *Hypnea* are quite common in the warm seas of the world.

Similar Species: *Bonnemaisonia hamifera* (plate 46) also bears hooks, but these are not at the ends of branches and the plant is rose-red and of a more delicate construction.

Range: Tropics to Southern Massachusetts.

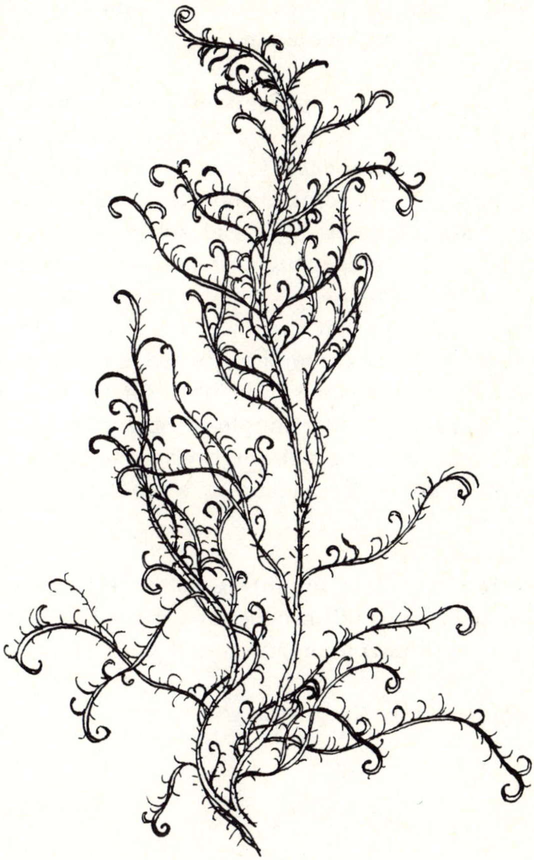

Plate 45. Hypnea musciformis

Bonnemaisonia hamifera

Bonnemaisonia hamifera	**bún - may - zó - nee - a** **ham - íf - er - a**
bonnemaisonia: after Bonnemaison, **a French botanist**	**hamifera: bearing** **hooks**

This is a very beautiful seaweed found during the summer months. It is a delicate, rose-red, finely branched plant which bears scattered, fleshy, hooklike branches. It grows in the sublittoral out into deep water and the beachcomber collector usually finds it washed ashore as a small fragment. When attached it grows in a bushy tangled mass only 3-4 inches high. It makes an excellent pressed specimen.

It has been discovered that a species long described as *Trailliella intricata* which grows as a compact, tufted red filament on other algae and rocks is probably one of the phases in the life cycle of *Bonnemaisonia*. This is just one example of how a seaweed may have several totally different stages through which it passes in its reproductive process. Many of these complex reproductive processes have not yet been traced by scientists.

Similar Species: *Hypnea musciformis* (plate 45) also bears hooks, but these are at the ends of branches and it is a larger, less delicate plant. It is also dull purple or purple-red.

Range: Long Island to Nova Scotia.

Plate 46. Bonnemaisonia hamifera

Nemalion helminthoides

Nemalion helminthoides	**nem - áy - lee - on hell - min - thóy - deez**
nemalion: crop of strings	**helminthoides: like a worm**
Common Name: "threadweed"	

These sparsely branched, slimy, wormlike strands are quite soft and do not stand erect but tend to hang downward over rocks in the intertidal zone, forming dark brown slippery colonies. Individual plants develop from a small disc and are 2-8 inches long, cylindrical, solid and somewhat elastic. They can be found during the summer, particularly in coastal areas which are subjected to heavy wave action.

Similar Species: none.

Range: Long Island to Nova Scotia.

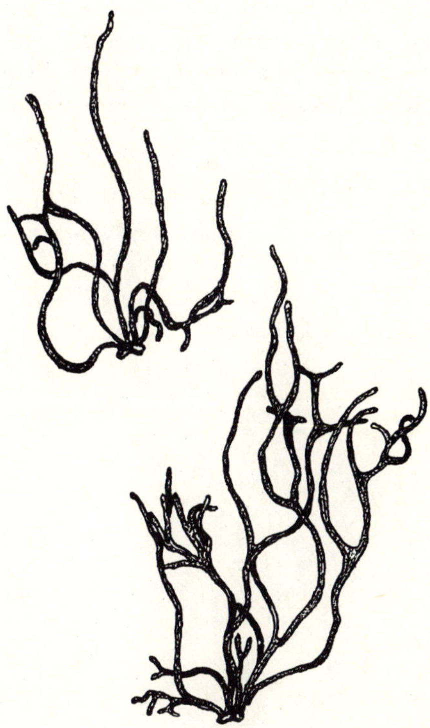

Plate 47. Nemalion helminthoides

Eudesme virescens

Eudesme virescens **you - déz - me vir - éss - ens**
eudesme: a good ligament **virescens: greenish**

This seaweed is of a peculiar gelatinous, slimy consistency. It develops on stones or other seaweeds as a somewhat hollow axis about 1/4 inch in diameter which branches only a few times, the terminal ones almost at right angles, and is between 4 inches and 1 foot in length. When examined closely the inner layers of tissue of this pale brown to olive plant can be seen as a somewhat distinct core under the soft outer tissue. It reproduces in the early summer and at that time it is mottled with dark brown spots.

Similar Species: *Nemalion* (plate 47) is also sparsely branched and slimy but is much darker, thinner, and does not grow on other algae.

Range: Long Island to Labrador.

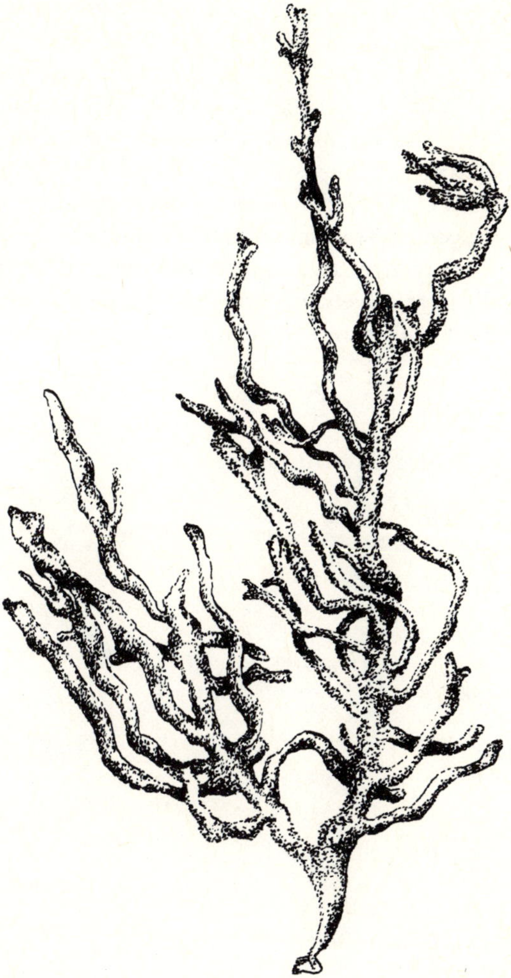

Plate 48. Eudesme virescens

Dasya baillouviana

Dasya baillouviana	**dáy - zee - a bay - léw - vee - á - na**
dasya: hairy	**baillouviana: after M. de Baillou, an 18th-century patron**

Common Name: "chenille weed"

Dasya, particularly when seen underwater, is perhaps the most beautiful of all the seaweeds discussed here. Its solid, cylindrical, sparingly branched axis is clothed with numerous delicate pink-red hairs. *Dasya* is a plant of the warm summer months and during that time there may occasionally be small dark reproductive structures on the branches. It may grow to a length of 2 feet, at depths to at least 40 feet, usually in relatively protected areas where there is good tidal flow. The plant makes an excellent pressed specimen.

Similar Species: none.

Range: Florida to Prince Edward Island.

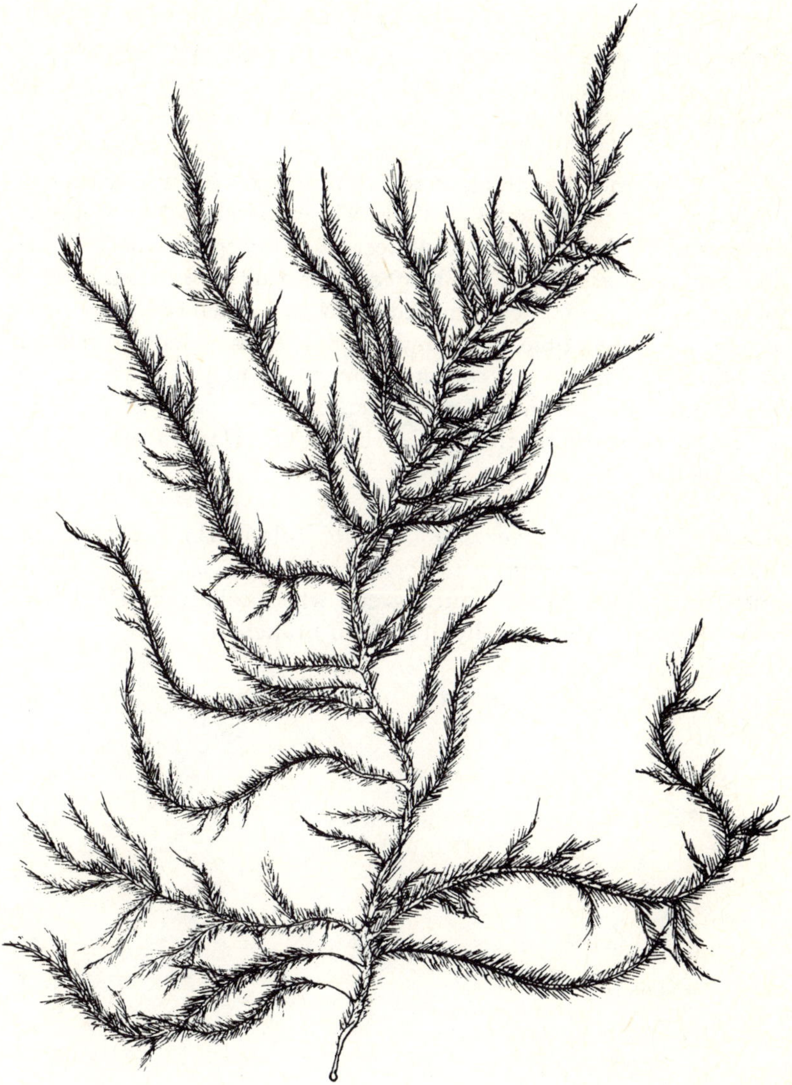

Plate 49. Dasya baillouviana

Desmarestia viridis

Desmarestia viridis **déz - ma - rés - ti - a vée - ree - dis**

desmarestia: after Desmarest, a French naturalist **viridis: green**

Common Names: "sea sorrel," "mermaid's hair"

Note the opposite branching on this common species. It is at its best in spring and early summer and grows in moderately deep water on rocks or shells down to at least 40 feet. Healthy specimens are light brown to olive-green and are 1-2 feet long. Collectors should watch for this species, for once removed from the water the plant soon degenerates into a blue-green highly acidic mass which will ruin all the other seaweeds in the collecting bucket or bag. Apparently the plant cells break down and release sulfuric and other acids which are highly corrosive, so this species should be isolated from the rest of the collection.

Similar Species: *Desmarestia aculeata* (plate 41) has alternate branching and is spiny. *Dictyosiphon* (plate 67) has thinner branches and it does not turn green when collected. *Chordaria* (plate 51) has wider, usually undivided branches.

Range: New Jersey to the Arctic.

Plate 50. Desmarestia viridis

Chordaria flagelliformis

Chordaria flagelliformis	**kor - dáy - ree - a fla - jéll - lee - fór - mis**
chordaria: cordlike	**flagelliformis: whiplike**

Both the main axis and the branches of this plant are slippery, dark brown and cordlike, and it has a firm, somewhat elastic consistency. Note that the branches toward the end of the plant are themselves unbranched, and are longer than the main axis, but about the same in diameter. *Chordaria* is a common intertidal plant up to 2 feet long. It is often found unattached and floating.

Similar Species: *Dictyosiphon* (plate 67) has more abundant, finer branches. *Desmarestia aculeata* (plate 41) has alternate branches and is spiny. *Desmarestia viridis* (plate 50) has abundant opposite branching.

Range: New Jersey to the Arctic.

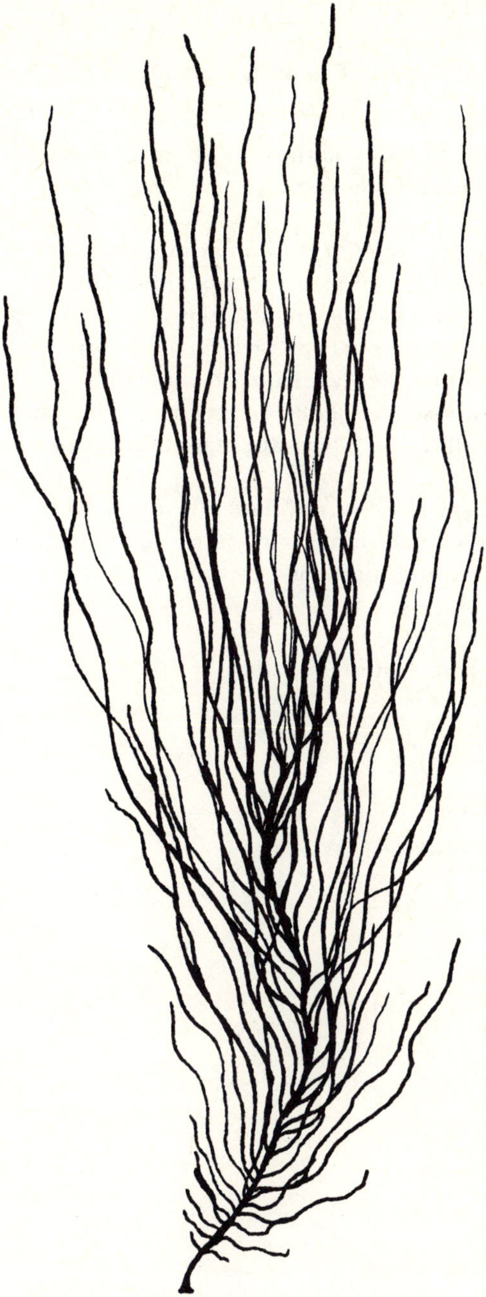

Plate 51. Chordaria flagelliformis

Champia parvula

Champia parvula **chám - pee - a pár - view - la**
champia: after M. Deschamps, a French botanist **parvula: small**

Flourishing during the warm summer months in the shallow sublittoral down to at least 30 feet, *Champia* often attaches to larger seaweeds and appears as a bushy, coarsely branching, dull red-brown plant. It has a peculiar "crispy" gelatinous texture, and when examined closely can be clearly seen to be made up of a series of slightly constricted cask-shaped hollow segments. The plant is only 1-3 inches long with rather short alternate tapered branches.

Similar Species: *Ceramium* (plate 66) may have a segmented appearance but is much thinner and more abundantly branched. The tips of the *Champia* branches are blunt, and not in the form of *Ceramium*'s pincers.

Range: Florida to Eastern Canada.

Plate 52. Champia parvula

Cystoclonium purpureum

Cystoclonium purpureum **sís - toe - kló - nee - um**
pur - púr - ee - um

cystoclonium: bladdery branches **purpureum: purple**

A very common summertime plant, *Cystoclonium* should be easily recognized. It has a main axis which is wider (1/8 inch) than the branches. The latter are subdivided 3-5 times and they are narrow at the base and taper to a very fine point. The plant has a characteristic firm fleshy texture. It grows from the lower intertidal zone down to at least 45 feet and is a rose-red to deep purple. There is an interesting variety occasionally found, the variety *cirrhosum,* which means "full of hair curls." In such a plant many of the branches become spirally coiled. The term "bladdery branches" refers to the fact that when the plant is in the reproductive stage, the development of small structures called cystocarps causes lumpy swellings among the branches. In deep water *Cystoclonium* is often found growing in large mats associated with *Phyllophora* (plate 27).

Similar Species: *Gloiosiphonia* (not illustrated) is not common. It resembles *Cystoclonium* except that there are several main axes arising from a common base, and these are tapered at both ends. The main branches are conspicuously tapered at the base. This plant may be found in the spring just below low tide from Long Island to Newfoundland.

Range: New Jersey to Newfoundland.

Plate 53. Cystoclonium purpureum

Agardhiella subulata

Agardhiella subulata **a - gár - dee - él - la súb - u - láy - ta**
agardhiella: after C. Agardh, a Swedish botanist **subulata: like an awl**

This seaweed is a bushy plant up to 1 foot long which has a firm fleshy consistency. It varies in color from a rose-red to pink and the entire plant body has a peculiar milky translucent character to it. It grows attached to stones or shells in relatively quiet water below low tide to a depth of at least 15 feet. Note that the main axis and branches are cylindrical and that the branches are tapered at both ends.

Similar Species: *Gracilaria* (plate 26) is flattened and the branches are not tapered at both ends.

Range: Tropics to Cape Cod, occasionally above to New Hampshire.

Plate 54. Agardhiella subulata

Lomentaria baileyana

Lomentaria baileyana	**lów - men - táy - ree - a** **báy - lee - án - na**
lomentaria: mealy	**baileyana: after Bailey,** **a New York botanist**

This seaweed grows in tufts 2-5 inches tall, on rocks, shells and larger seaweeds in relatively quiet water to at least 30 feet. It has a soft texture and is purple-red to pink. The branching pattern is a good clue to the species because the cylindrical branches are irregularly alternate, and the smallest branches are curved and tapered at both ends, and often appear to all arise from only one side of the branch.

Similar Species: *Chondria* (plate 56) branches are shorter and not curved.

Range: Florida to Nova Scotia.

Plate 55. Lomentaria baileyana

Chondria sp.

Chondria sp. **kón - dree - a**
chondria: cartilagelike

There are several species of *Chondria* within our range, and they can be difficult to separate. *Chondria* is a rather soft, branching plant, about 3-9 inches tall with an erect, very thin main axis. The numerous alternate branches are short and tapered at both ends. The plant may be a dull purple, but is often a pale straw color, especially in shallow water. *Chondria baileyana* is often found attached to species of *Fucus* (plates 18, 20, 21) in sheltered coves, while other species grow just below the low tide line in more open water.

Similar Species: *Lomentaria* (plate 55) has longer branches which are curved.

Range: Tropics to Eastern Canada.

Plate 56. Chondria sp.

KEY TO VERY THIN, BRANCHED SEAWEEDS

1. Plant green	2
1. Not as above	5
2. Small highly branched segments roughly triangular in outline	*Bryopsis plumosa* (plate 57)
2. Not as above	3
3. Large, filamentous entangled branched mass	*Cladophora* sp. (plate 58)
3. Not as above	4
4. Often as small tuft, twisted together at base	*Spongomorpha arcta* (plate 59)
4. Very fine masses of sparsely branched filaments	*Rhizoclonium* sp. (plate 60)
5. Very small brown filaments attached to *Fucus or Ascophyllum*	6
5. Not as above	7
6. Attached to *Fucus*	*Elachista fucicola* (plate 61)
6. Attached to *Ascophyllum*	*Polysiphonia lanosa* (plate 62)
7. Golden brown cloudy masses of very fine filaments	*Ectocarpus* sp. (plate 63)
7. Not as above	8
8. Dark red velvety tufts on intertidal rocks	*Audouinella purpurea* (plate 64)
8. Not as above	9
9. Spherical masses of very fine red filaments (sphere about golf ball size)	*Spermothamnion* sp. (plate 65)
9. Not as above	10
10. Bushy, filamentous, definite banded appearance. Often ending in tiny Y-shaped "pincers"	*Ceramium* sp. (plate 66)
10. No Y-shaped "pincers"	11

11. Bushy, filamentous, made up of tiny elongated tubular cells	*Polysiphonia* sp. (plate 62)
11. Not as above	12
12. Large (up to 2-1/2 feet) bushy plant, very thin axis and alternate branches. Often on *Chordaria*	*Dictyosiphon foeniculaceus* (plate 67)
12. Not as above	13
13. Bushy, somewhat wiry, in winter and spring, soft tufts of small branches toward ends of plant	*Rhodomela confervoides* (plate 68)
13. Delicate, small red tufts	*Callithamnion* sp. or *Antithamnion* sp. (plates 69, 70)

Bryopsis plumosa

Bryopsis plumosa **bry - óp - sis plew - mó - sa**
bryopsis: mosslike **plumosa: feathery**
Common Name: "Christmas tree"

During the summer and autumn months, if you swim along the side of a jetty or natural rock formation you may find the delicate, light green fronds of this curious plant. Its featherlike branching pattern causes it to appear as a flat, triangular subdivided blade, with a distinct main axis. It is one of the few seaweeds in our area which is a "siphon," that is, the entire plant lacks internal cross walls and is therefore one very large simple cell. Under a microscope each tiny branch can be seen to be constricted at the base, and if reproductive, the plant could be seen to be either male or female. *Bryopsis* is not easy to find, but where it is able to grow it may be found in abundance.

Similar Species: *Plumaria* (plate 29) and *Ptilota* (plate 30) are also feathery but both are red and firm, not green and delicate.

Range: Florida to Nova Scotia.

Plate 57. Bryopsis plumosa

Cladophora sp.

Cladophora sp. **clad - óff - or -a**
cladophora: bearing branches

As the name suggests, *Cladophora* is profusely branched, but these branches are so thin that the whole plant, which may sometimes reach several feet in length, has a delicate, almost cloudy appearance when seen underwater. It is usually rather soft in texture, and when detached from its substrate in depths of up to 50 feet or more the plants may slowly tumble along the ocean floor and form spherical masses. Certain *Cladophora* species grow in fresh water and there they often become entangled in the shape of these so-called "*Cladophora* balls." The various species of *Cladophora,* of which there are about 12 in our area (and up to 400 worldwide), are difficult to separate and their differences are principally microscopic characteristics.

Similar Species: *Spongomorpha* (plate 59) and *Rhizoclonium* (plate 60) are finely filamentous green seaweeds, but the former is often coarse, and is shorter and tufted, while the latter is sparingly branched if at all.

Range: Florida to Labrador.

Plate 58. Cladophora sp.

Spongomorpha arcta

Spongomorpha arcta **spúng - o - mór - fa árc - ta**
spongomorpha: sponge-shaped **arcta: narrow**

The genus *Spongomorpha,* represented by 4 species in our area, grows as a small tuft or hemispherical mass of abundantly branched, uniseriate, bright green filaments. The plants tend to be twisted and held together at their bases so that they form ropelike masses topped by less matted branches. They attach to other seaweeds and rocks in the intertidal zone and shallow sublittoral.

Similar Species: *Cladophora* (plate 58) is branched but can be a much larger plant and is not twisted together at the base. *Rhizoclonium* (plate 60) is sparingly branched if at all.

Range: New Jersey to Newfoundland.

Plate 59. Spongomorpha arcta

Rhizoclonium sp.

Rhizoclonium sp. **ríse - o - kló - nee - um**
rhizoclonium: a root branch

This is another one of those seaweeds in which the identification of particular species can only be accomplished with a microscope. The genus *Rhizoclonium* grows as a mass of dark to light green very fine entangled filaments, which are only sparingly branched if at all. It grows in the intertidal zone, sometimes forming extensive mats on the rocks, especially in tide pools.

Similar Species: *Cladophora* (plate 58) and *Spongomorpha* (plate 59) are abundantly branched. *Chaetomorpha linum* (plate 35) is much coarser.

Range: North Carolina to Baffin Island.

Plate 60. Rhizoclonium sp.

Elachista fucicola

Elachista fucicola	**él - a - kíss - ta few - síck - o - la**
elachista: the least	**fucicola: a little seaweed**

The habitat is the clue to this species. *Elachista* grows as ½ inch tufts of brown uniseriate filaments scattered along the fronds of *Fucus* and occasionally other large seaweeds. The bases of the filaments penetrate the tissue of the host and form a firm cushionlike "button" which can be felt when the specimen is handled. This plant is often abundant in the lower intertidal in the summer and is inconspicuous during the winter months.

Similar Species: *Sphacelaria* (not illustrated) is quite similar but lacks the basal "button." Some invertebrates may look like tufted algae, but are not brown and soft to the touch.

Range: Virginia to Ellesmere Island.

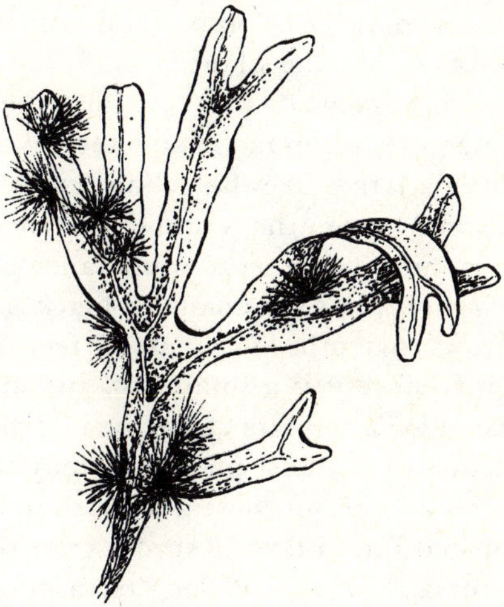

Plate 61. Elachista fucicola

Polysiphonia sp.

Polysiphonia sp. **póll - lee - sigh - fó - nee - a**
polysiphonia: many tubes

There are about 200 species of *Polysiphonia* world-wide, and at least 12 within our range. As in the case of *Ceramium* (plate 66) and others, species distinctions are generally based upon microscopic characteristics. Two common species are illustrated here, *Polysiphonia nigrescens* and *Polysiphonia lanosa*. Specimens of the genus are easily recognized first of all by the fact that they are very fine, abundantly branched plants that can really be confused only with *Ceramium* species (plate 66). Close examination (a hand lens is very useful here) reveals that the plant is made up of a series of tubes wrapped around a central axis to form cylindrical axes and branches. These tubes are in fact elongated cylindrical "pericentral" cells surrounding a large, likewise cylindrical "central" cell. Species differences depend upon the number of these pericentral cells around the axis, and the general appearance of the plant.

Polysiphonia nigrescens is a common black seaweed, up to 1 foot long, which grows to depths of at least 50 feet. *Polysiphonia lanosa* is easy to identify since it is a small wiry tufted plant which grows attached to *Ascophyllum* (plate 22) in the intertidal zone. It actually wounds the tissues of the host plant and may stunt its growth.

Polysiphonia is another of those genera of red seaweeds that have a very complicated life history. Reproductive plants may be either male, female, or a third type called "tetrasporic." All three plants look very much alike, even though they are quite different functionally. Even the "lowly" seaweeds may have more complicated reproductive processes than many of the "higher" organisms, including man.

Similar Species: *Ceramium* (plate 66) has a banded appearance and has pincerlike endings.

Range: Florida to the Arctic.

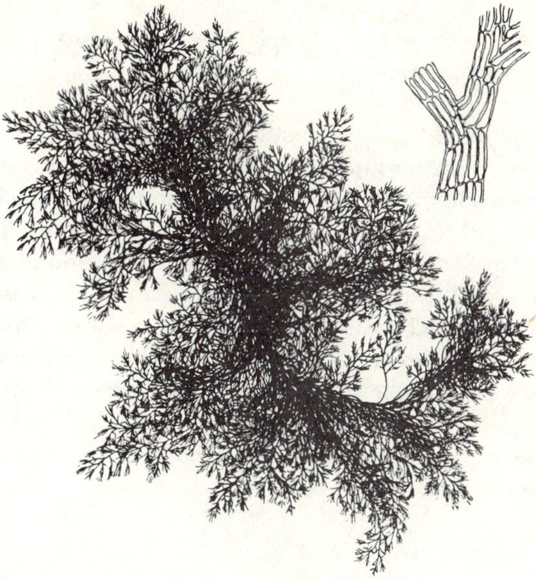

Plate 62 (a). Polysiphonia nigrescens
Plate 62 (b). Polysiphonia lanosa

Ectocarpus sp.

Ectocarpus sp. **ék - toe - cár - pus**
ectocarpus: external fruit

Plants of the genus *Ectocarpus* should be easy for the diver to recognize because they appear almost like yellow-brown clouds suspended above substrates such as stones, shells or coarser algal species. Some species are somewhat twisted so that a main axis can be discerned. Mature plants are about 1 foot long, and grow year round down to a depth of at least 40 feet. Under magnification the plant is revealed as many highly branched uniseriate filaments which often end in long colorless hairs. Species differences depend upon microscopic characteristics such as the shape of reproductive structures.

Similar Species: *Ceramium* (plate 66) and *Polysiphonia* (plate 62) are darker and coarser.

Range: Florida to Devon Island.

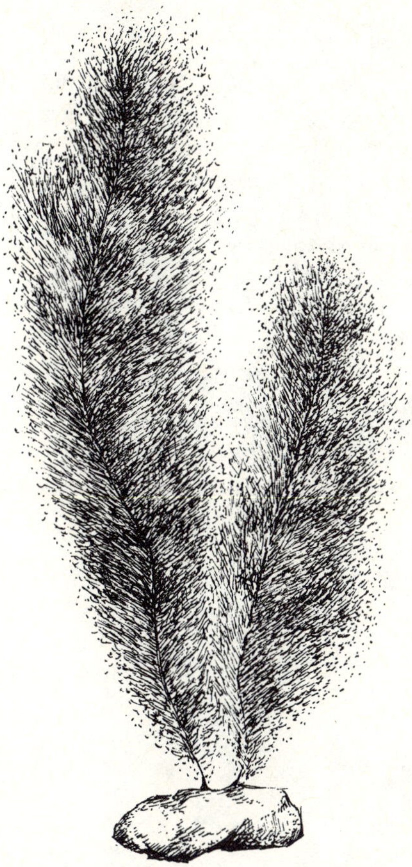

Plate 63. Ectocarpus sp.

Audouinella purpurea

Audouinella purpurea — **awe - dó - in - éll - a pur - púr - ee - a**

audouinella: after J. V. Audouin, a French entomologist

purpurea: purple

When growing sparingly *Audouinella* is an inconspicuous seaweed, but when in abundance it may cover shaded rocks near the low tide line with a beautiful dark purple-red slippery "turf." Microscopically, it consists of filaments which creep over the substrate for attachment, and sparsely branched erect uniseriate photosynthetic filaments, about 1/2 inch long. Look for it by moving aside the *Fucus* (plates 18, 20, 21) or *Ascophyllum* (plate 22) draped over the rocks, or examine the surfaces of pilings or the base of *Laminaria* species (plates 7, 11-13).

Similar Species: *Audouinella spetsbergensis* (not illustrated) is quite similar but less common, sometimes on *Chaetomorpha* (plate 36) or *Phyllophora* (plate 27).

Range: Long Island to Ellesmere Island.

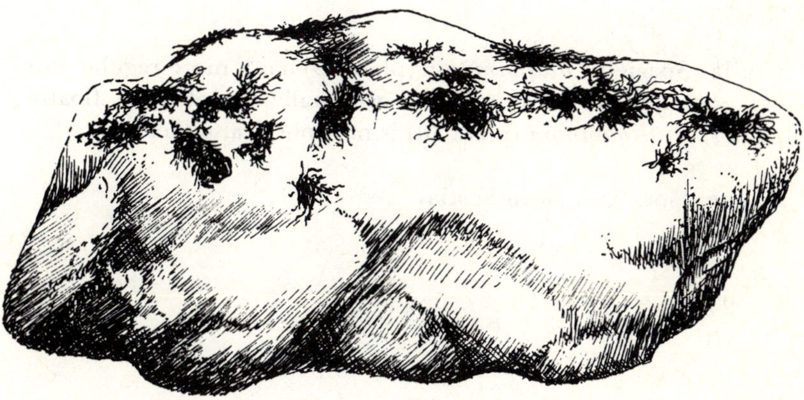

Plate 64. Audouinella purpurea

Spermothamnion sp.

Spermothamnion sp. **spérm - o - thám - nee - on**
spermothamnion: seed shrub

If you find a "fuzzy" clump of bright rosy red extremely delicate branched uniseriate filaments, about the size of a golf ball, you probably have found *Spermothamnion*. It grows attached to *Chondrus* (plate 33), *Phyllophora* (plate 27) and other seaweeds down to at least 40 feet, but commonly is seen detached and floating. Its beautiful color and texture can only be fully appreciated underwater, since it collapses into a soggy lump when removed from the water.

Similar Species: *Callithamnion* (plate 69) has a more regular forking branching pattern, and is not usually in spherical floating masses. *Antithamnion* (plate 70) is not spherical either.

Range: Florida to Nova Scotia.

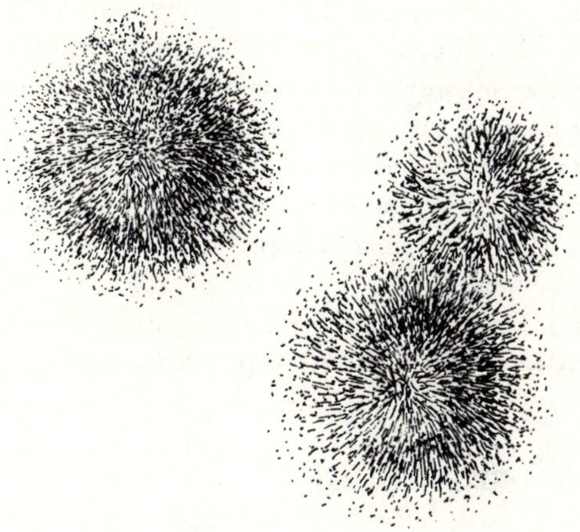

Plate 65. Spermothamnion sp.

Ceramium sp.

Ceramium sp. **sir - ráy - me - um**
ceramium: a pitcher
Common Name: "pitcher weed"

It is hardly possible to look for seaweeds and not come across a species of *Ceramium* at any time of the year. The differences between our local species are based upon microscopic characteristics, but the genus *Ceramium* should be easily recognized. It is a bushy, dark red, abundantly branched plant which, upon close examination, exhibits a distinct "banded" appearance. These "bands" are actually layers of tiny cells covering the central axis which is a uniseriate row of large cylindrical cells. Also, the branch tips look like pincers or, as some say, like lobster claws. When reproductive, small cuplike structures may be seen among the branches. A hand lens is useful for viewing these characteristics more closely. *Ceramium* grows in the intertidal zone and out to moderate depths on many substrates, and is commonly seen unattached and drifting.

Similar Species: *Polysiphonia* (plate 62) is quite similar in general appearance but lacks the banded appearance and the pincer endings.

Range: Tropics to Baffin Island.

Plate 66. Ceramium sp.

Dictyosiphon foeniculaceus

Dictyosiphon foeniculaceus	**díc - tea - o - sígh - fon** **fo - níc - u - láy - shus**
dictyosiphon: a netted tube	**foeniculaceus: like fennel, an herb**

The very thin conspicuous main axis and alternate branching help to distinguish this seaweed from all others. It may reach a length of 2 1/2 feet and appear quite bushy. Young plants are yellow-brown but they darken to a black with age. *Dictyosiphon* is often found attached to *Chordaria* (plate 51) and this can be confusing as the plants are somewhat similar in appearance. It is present all year round but is in its best condition during the summer. With careful separation of the branches it makes a fine pressed specimen.

Similar Species: *Chordaria* (plate 51) has wider, usually undivided branches. *Desmarestia viridis* (plate 50) has opposite branching and turns green when removed from the water. Old *Desmarestia aculeata* (plate 41) has numerous small spines left when the deciduous hairs fall off.

Range: New Jersey to Nova Scotia.

Plate 67. Dictyosiphon foeniculaceus

Rhodomela confervoides

Rhodomela confervoides	**róe - doe - mé - la** **kón - fur - vóy - deez**
rhodomela: red-black	**confervoides: like a water plant**

This is a somewhat wiry, red to black bushy seaweed which may be 16 inches tall. In the winter and spring it is relatively easy to identify because of the tufts of small branches which develop toward the outer parts of the plant. In the summer these tufts fall off and the plant is quite ragged in appearance. It can be found growing on stones in relatively shallow water to at least 10 feet depth.

Similar Species: Several other species of *Rhodomela* are found in our area but are very difficult to distinguish from the above. Old specimens of *Dictyosiphon* (plate 67) may be confused with this but *Rhodomela* has fewer branches off the main axis.

Range: New Jersey to Baffin Island.

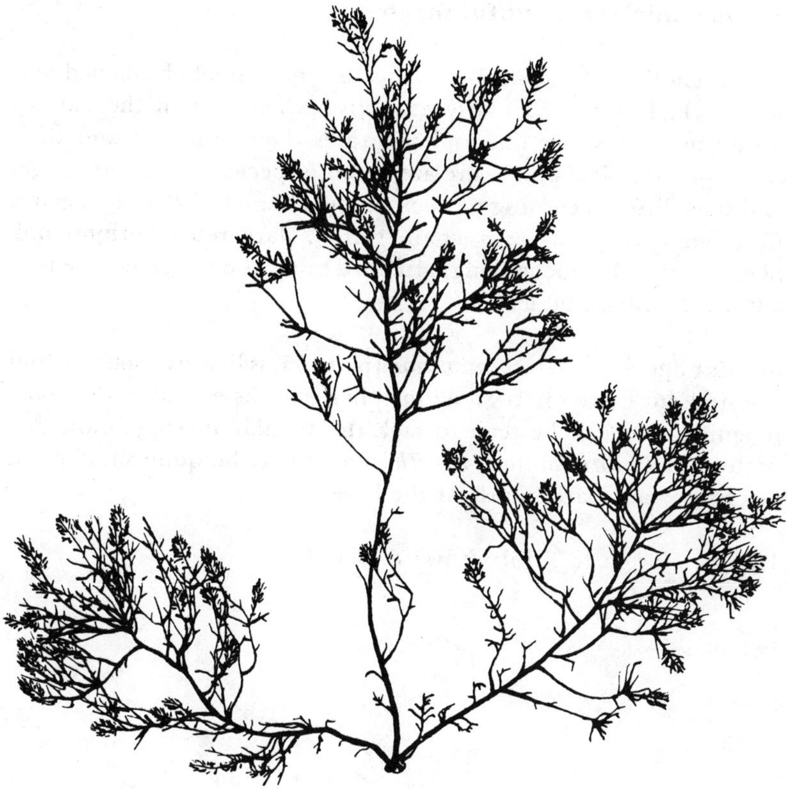

Plate 68. Rhodomela confervoides

Callithamnion sp.

Callithamnion sp. **kál - lee - thám - nee - on**
callithamnion: a beautiful shrub

This is another of those rather delicate, small, finely branched seaweeds which are particularly attractive when seen in the natural state when the water lifts and separates their branches and their coloration is at its best. There are about 5 species within our range, and they differ according to microscopic characteristics. The genus *Callithamnion* grows as a small, tufted, dark red to bright pink plant, usually 1-3 inches long, attached to rocks or large seaweeds at depths to at least 50 feet.

Similar Species: *Spermothamnion* (plate 65) is likewise delicate and rose-red but more often grows in spherical masses and under some magnification can be seen to lack the regular forking branching pattern of *Callithamnion*. *Antithamnion* may be quite similar but has finer axes, particularly at the base.

Range: Florida to Prince Edward Island.

Plate 69. Callithamnion sp.

Antithamnion sp.

Antithamnion sp. **án - ti - thám - nee - on**
antithamnion: opposite shrub

There are 6 species of the genus *Antithamnion* within our range, and again, species differences are based on microscopic characteristics that do not concern us here. All of the species are remarkably delicate and consist of very thin, uniseriate branched filaments. They may be found attached to stones or larger algae from tide pools down to at least 100 feet. The largest specimens get to be about 8 inches long, but they are commonly only 2-3 inches. Sometimes *Antithamnion* can be recognized by the fact that in some species the filaments toward the ends of the plant (see illustration) are tiny and clustered and appear as bushy "spots." The plant is a beautiful rose-red to rose-pink.

Similar Species: *Spermothamnion* (plate 65) grows in more spherical masses. *Callithamnion* (plate 69) is coarser at the base and lacks bushy "spots."

Range: Virginia to Baffin Island.

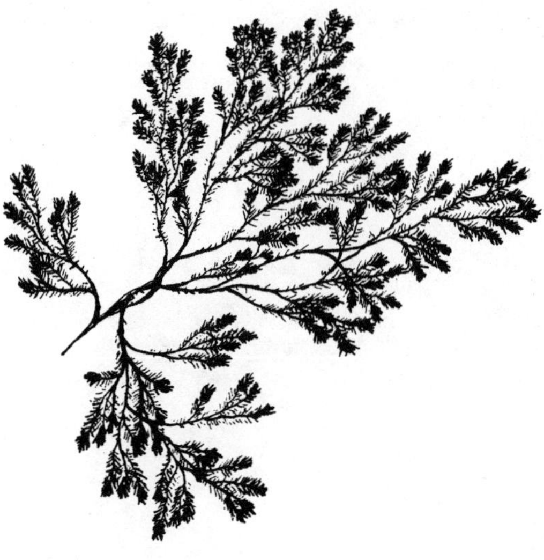

Plate 70. Antithamnion sp.

KEY TO CRUSTS ON ROCKS AND SHELLS

1. Dark slimy layer above barnacle zone, high in the intertidal	*Calothrix* sp. (plate 71)
1. Not as above	2
2. Dark brown to black, thin to thick leathery crust, sometimes brittle, or with concentric markings	*Ralfsia* sp. (plate 72)
2. Not as above	3
3. Dark red to brown-red, very thin irregular patches	*Hildenbrandia rubra* (plate 73)
3. Not as above	4
4. Smooth, curved ridges on thick purple to yellow crust	*Phymatolithon* sp. (plate 74)
4. Not as above	5
5. Smooth, thin pink crust	*Clathromorphum* sp. (plate 74)
5. Many irregular protuberances from surface of pink-red crust	*Lithothamnion* sp. (plate 74)

Calothrix sp.

Calothrix sp. **kál - o - thrix**
calothrix: beautiful hair

This is the only microscopic alga treated in this book. The individual plants of *Calothrix* are filaments only 1/10 inch long, but *en masse* they form conspicuous blue-green to black slimy, slippery patches on rocks in the upper and middle intertidal. The colony of plants is closely adherent to the rock and somewhat resembles a stripe of dark paint. This is the alga that may form the dark band above the barnacle zone discussed in the Introduction (p. 27). *Calothrix* is a member of the Division Cyanophyta, the "blue-green" algae. It is highly resistant to drying out and hence can live in the most often exposed areas of the intertidal. *Calothrix,* like some of the other members of the Cyanophyta, has the capacity to extract nitrogen gas from the air and incorporate it into proteins. This ability to "fix" nitrogen is rare among organisms and is vital in supplying nitrogen in a usable form to the rest of the living world.

Similar Species: *Verrucaria* (not illustrated) which is a lichen, a combination of algal and fungal cells, forms a black crust which may also be found in the intertidal but is usually formed into rounded, discrete patches somewhat thicker than *Calothrix*. It resembles black tar patches on the rocks.

Range: Tropics to the Arctic.

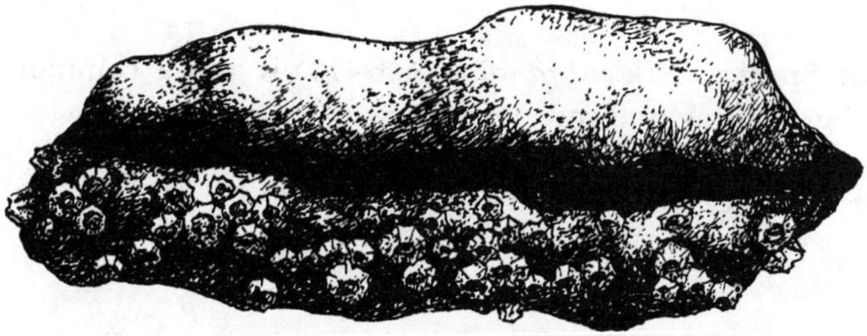

Plate 71. Calothrix sp.

Ralfsia sp.

Ralfsia sp. **rálf - see - a**
ralfsia: after John Ralfs, Esq., a British botanist

This crust-forming genus is represented by several species which may be quite unlike each other in appearance. A common one is *Ralfsia verrucosa,* which grows as a dark brown leathery crust on rocks and shells in the intertidal out to depths of at least 50 feet. Older specimens become loose and brittle and the edges of the plants are often upturned. *Ralfsia fungiformis* looks just like a wood-rotting bracket fungus with its rounded, overlapping lobes which are brown to black in color and are marked with concentric rings.

Similar Species: *Hildenbrandia* (plate 73) is also a common thin crust, but is slimy and more reddish.

Range: New Jersey to the Arctic.

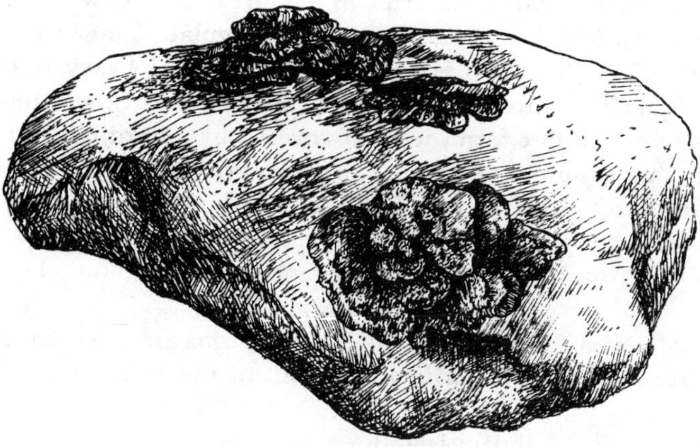

Plate 72. Ralfsia fungiformis

Hildenbrandia rubra

Hildenbrandia rubra	**híll - den - brán - dy - a róo - bra**
hildenbrandia: after F. Hildenbrandt	**rubra: reddish**

The observant diver cannot help but be aware that the rocky ocean bottom is covered with a beautiful mosaic of varied hues of pinks, purples and reds, and that even in areas where there are relatively few large seaweeds these crustlike forms grow together in a bewildering array of patterns and surface textures. *Hildenbrandia* is the dominant non-calcified red crust in our area, and appears in this mosaic as dark blood-red or brown-red irregular, somewhat slimy patches. Their thickness is about that of several heavy coats of paint. By non-calcified we mean that unlike most of the other accompanying crustose forms it is not covered by a hard layer of calcium or magnesium carbonate. The plants grow from the low intertidal down to at least 70 feet. It is most easily collected intact, like other crustose specimens, by simply picking up a small rock on which it is growing. It will retain its appearance indefinitely.

Similar Species: Most other local crustose forms are calcified. *Ralfsia* (plate 72) is not, but it is brown and may be thick and brittle.

Range: Florida to Baffin Island.

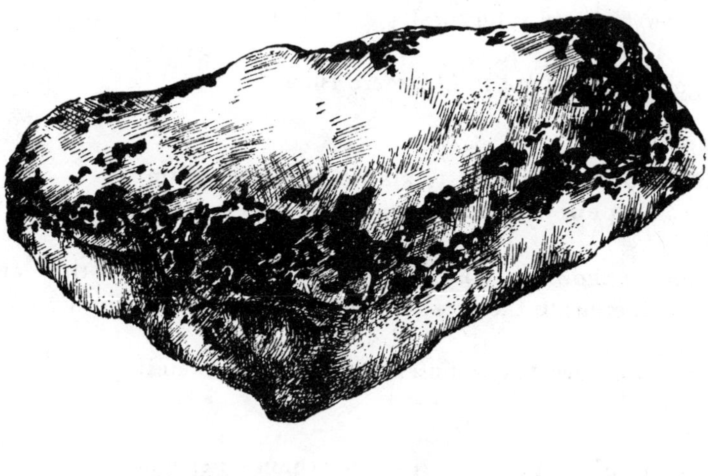

Plate 73. Hildenbrandia rubra

Calcified Crustose Red Algae

These are hard plants which are encrusted with a stony covering of calcium and magnesium carbonate. They are very common but difficult to identify to species or even genus without specialized techniques. Below are three common species:

Phymatolithon sp. **fy - mát - o - líth - on**
phymatolithon: a stone swelling

The species illustrated is marked by smooth curved ridges on the thick crust. The surface is very smooth and light purple to yellowish. It is intertidal and sublittoral.

Range: Rhode Island to Ellesmere Island.

Clathromorphum sp. **kláth - ro - mór - fum**
clathromorphum: lattice-shaped

The species shown here is light pink and smooth, rather thin, and tightly adherent to the rock. It is intertidal and sublittoral.

Range: Northern Massachusetts to Ellesmere Island.

Lithothamnion sp. **lí - tho - thám - nee - on**
lithothamnion: stone shrub

This species is characterized by hard nipplelike protuberances from the underlying pink-red crust. It is sublittoral.

Range: New Jersey to the Arctic.

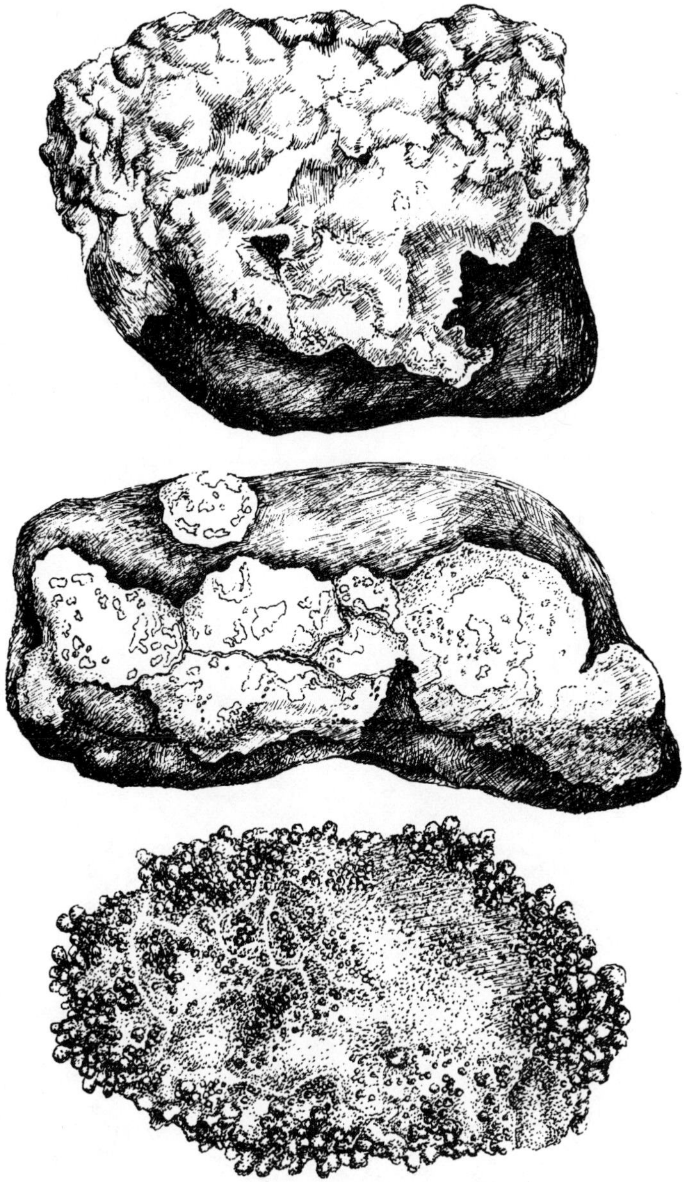

Plate 74 (a). Phymatolithon sp.
Plate 74 (b). Clathromorphum sp.
Plate 74 (c). Lithothamnion sp.

KEY TO "MISCELLANEOUS" SEAWEEDS

1. Firm, "crumpled spheres," golden-brown, often on *Corallina*	*Leathesia difformis* (plate 75)
1. Plant elaborate, brown, with main axis, pointed leaflike blades, small air bladders on short stalks	*Sargassum filipendula* (plate 76)

Leathesia difformis

Leathesia difformis — **lee - thée - zi - a dif - fór - mis**
leathesia: after G.R. Leathes, a British naturalist — **difformis: shaped double**
Common Names: "sea potato," "rat's brains"

The common names of this seaweed are really quite descriptive of its appearance. It is a hollow, yellow-brown crumpled sphere about the size of a golf ball, and usually grows attached to *Corallina* (plate 17) or *Chondrus* (plate 33). When sectioned and viewed under a microscope, *Leathesia* can be seen to be simply a compact mass of highly branched filaments, colorless toward the inside of the plant and filled with photosynthetic pigments toward the surface. There is not much point in collecting *Leathesia,* since it can't be pressed, but look for it during the summer months scattered throughout beds of *Corallina* and "Irish moss."

Similar Species: none.

Range: North Carolina to Nova Scotia.

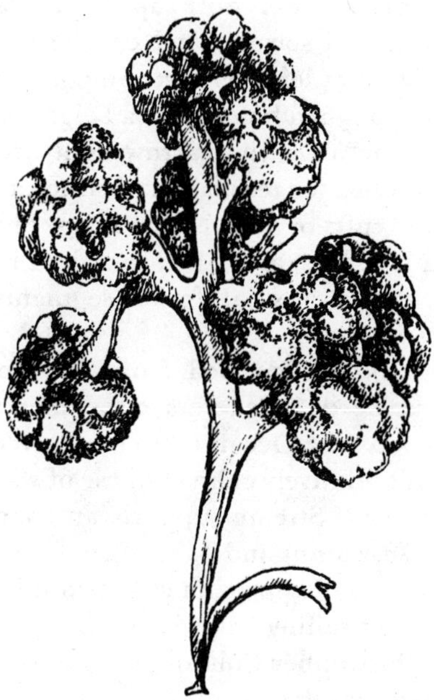

Plate 75. Leathesia difformis

Sargassum filipendula

Sargassum filipendula — **sar - gás - um fíll - lee - pén - du - la**

sargassum: floating seaweed — **filipendula: hanging thread**

This is a beautifully elaborate seaweed that one could almost picture as being a typical land plant because it appears to have a stem, leaves, and even berries. However, as explained in the Introduction (p. 9), such structures are not found in the seaweeds and the resemblance, however remarkable, is only superficial. *Sargassum filipendula* is our only attached species, and can be found from shallow water out to a depth of at least 95 feet. Smooth, sparsely branched main axes arise from a lobed holdfast to a height of up to 2 feet, and toward the base of the plant the narrow light brown lance-shaped leaves have saw-toothed edges and a distinct midrib. They are covered by scattered tufts of dark brown hairs. Short stalks support small spherical air bladders, and when the plant is fertile the reproductive structures appear as small forked segments arising from the bases of the leaves.

The term *Sargassum* is derived from the Spanish "sargazo," meaning floating seaweed. This refers to the phenomenon occurring in that part of the Atlantic Ocean between the West Indies and the Azores, in which this relatively calm expanse of water surrounded by the rapidly flowing Gulf Stream supports an enormous population of floating *Sargassum natans* and *Sargassum fluitans.* This Sargasso Sea, as it is called, covers approximately two million square miles. Ancient legends tell of sailing ships becoming hopelessly entangled in the weed, and Christopher Columbus reported sailing through it on his first westward voyage.

The *Sargassum* plants in this enormous population reproduce not by sperm and eggs as does our local species, but simply by fragmenting into pieces which have the capacity to grow into adult plants. The tangled masses of *Sargassum* nourish and protect a wide variety of animal life, which in some cases has adopted protective coloration and body parts which quite closely resemble the seaweed.

Similar Species: *Sargassum natans* (not illustrated) may occasionally wash ashore from Florida to Massachusetts. It is native to the Caribbean and its air bladders are tipped with a spine. It does not bear reproductive structures and lacks the hairy tufts on the leaves.

Sargassum fluitans (not illustrated) may also wash in occasionally from the Gulf Stream. Its bladders are oval and the stalks are often winged with a ridge of tissue.

Range: Florida to Southern Massachusetts.

Plate 76. Sargassum filipendula

Bibliography

BOOKS—GENERAL

Bold, Harold C., and Michael J. Wynne. *Introduction to the Algae: Structure and Reproduction*. Prentice-Hall, Tarrytown, New York. 1978.
A detailed analysis which includes seaweeds along with the rest of the algae.

Chapman, A.R. *Biology of Seaweeds*. University Park, Baltimore, Maryland. 1979.
A short technical account of seaweed function.

Dawes, Clinton J. *Marine Botany*. Wiley, New York. 1981.
A good introductory textbook on the subject.

Dring, M.J. *The Biology of Marine Plants*. E. Arnold, London. 1982.
Another good introduction to these organisms.

Lobban, Christopher S., and Michael J. Wynne (eds.). *The Biology of Seaweeds*. University of California Press, Berkeley, California. 1982.
A comprehensive analysis of the functions of seaweeds.

Trainor, F.R. *Introductory Phycology*. Wiley, New York. 1978.
An excellent introduction to the algae including the seaweeds. A good start for the beginner.

SEAWEED GUIDES

Abbott, Isabella A., and G.J. Hollenberg. *Marine Algae of California*. Stanford University Press, Stanford, California. 1976.
A guide to those seaweeds, written with scientists in mind.

Abbott, Isabella A., and E. Yale Dawson. *How to Know the Seaweeds*. William C. Brown Co., Dubuque, Iowa. Second edition, 1978.
An illustrated key including Pacific seaweeds.

Arnold, Augusta F. *The Sea-Beach at Ebb Tide*. Dover Publications, Inc., New York. 1968.
An interesting reprint of a 1901 book covering plants and animals of the North American seacoast.

Dawes, Clinton J. *Marine Algae of the West Coast of Florida*. University of Miami Press, Baltimore, Maryland. 1970.

A useful guide to the seaweeds of that region which are in general accessible only to the diver because of the limited intertidal zone.

Dickinson, Carola I. *British Seaweeds*. Eyre and Spottiswoode, London. 1963.

Many seaweeds which grow in British waters also can be found along the East Coast of North America.

Duddington, C.L. *Beginner's Guide to Seaweeds*. Drake Publishers, New York. 1971.

Another British text, but much here applicable to our local area.

Guberlet, Muriel L. *Seaweeds at Ebb Tide*. University of Washington Press. 1956.

An illustrated key to Pacific seaweeds.

Hervey, A.B. *Sea Mosses*. Orange Judd Co., New York. 1882.

A delightful 19th-century treatise on British seaweeds. Hard to come by—available at Woods Hole Marine Biological Laboratories, Massachusetts.

Hillson, Charles J. *Seaweeds: A Color-Coded Guide to Common Marine Plants of the East Coast of the United States*. Keystone Books, New York. 1977.

Seaweeds arranged by color and written for the general reader.

Humm, Harold J. *The Marine Algae of Virginia*. University Press of Virginia, Charlottesville. 1979.

A comprehensive guide written for scientists on the algal flora of that region.

Kapraun, Donald F. *An Illustrated Guide to the Benthic Marine Algae of Coastal North Carolina*. I. *Rhodophyta*. The University of North Carolina Press, Chapel Hill, 1980.

A guide for scientists to the red algae of the above area.

Kapraun, Donald F. *An Illustrated Guide to the Benthic Marine Algae of Coastal North Carolina*. II. *Chlorophyta*. Lubrecht and Cramer, Forestburgh, New York. 1984.

As above, for the green algae.

Kingsbury, John M. *Seaweeds of Cape Cod and the Islands*. The Chatham Press Inc., Chatham, Massachusetts. 1969.

An excellent illustrated descriptive treatment of the more common seaweeds of that region, with useful references to specific localities.

Landsborough, Rev. D. *A Popular History of British Seaweeds*. Reeve and Benham, London. 1851.

Another interesting 19th-century volume. Also available at Woods Hole, Massachusetts.

Petry, Loren C. *A Beachcomber's Botany*. Chatham Conservation Foundation, Chatham, Massachusetts. 1968.

A well-illustrated guide to mainly coastal land plants but includes some seaweeds.

Robbins, Sarah F., and Clarice Yentsch. *The Sea Is All About Us*. The Peabody Museum of Salem and the Cape Ann Society for Marine Science, Inc. 1973.

A useful guide to plants and animals with particular reference to the Cape Ann, Massachusetts area.

Taylor, William Randolph. *Marine Algae of the Northeast Coast of North America*. University of Michigan Press, Ann Arbor. Second edition, 1957.

The most comprehensive technical treatment of the seaweeds of this area written to date—not a beginner's guide.

Waaland, J. Robert. *Common Seaweeds of the Pacific Coast.* Pacific Search Press/Nature, Seattle. 1977.

A handy guide for the layperson to common marine algae of that region.

SEAWEED USES

Chapman, V.J., and D.J. Chapman. *Seaweeds and Their Uses.* Chapman and Hall, New York. 1980.

A comprehensive treatment of seaweed utilization world-wide.

McConnaughey, Evelyn. *Sea Vegetables: Harvesting Guide and Cookbook.* Naturegraph, Happy Camp, California. 1985.

A handy amateur's guide to gathering and preparing seaweeds.

Madlener, Judith Cooper. *Sea Vegetable Gelatin Cookbook.* Woodbridge Press, Santa Barbara, California. 1981.

As the above.

Okazaki, Akio. *Seaweeds and Their Uses in Japan.* Tokai University Press. 1971.

A fascinating account of the multitude of uses to which seaweeds are put in the Orient.

Stephenson, W.A. *Seaweeds in Agriculture and Horticulture.* EP Publishing Ltd., East Ardsley, Wakefield, Yorkshire, England. 1973.

A very useful guide to those interested in using seaweeds in organic gardening.

Index to the Species

Division Cyanophyta ("blue-green" algae)

Division Chlorophyta ("green" algae)

Division Phaeophyta ("brown" algae)

A CATALOG OF SELECTED

DOVER BOOKS

IN ALL FIELDS OF INTEREST

A CATALOG OF SELECTED DOVER BOOKS IN ALL FIELDS OF INTEREST

25 KITES THAT FLY, Leslie Hunt. Full, easy-to-follow instructions for kites made from inexpensive materials. Many novelties. 70 illustrations. 110pp. 5⅜ × 8½. 22550-X Pa. $2.50

PIANO TUNING, J. Cree Fischer. Clearest, best book for beginner, amateur. Simple repairs, raising dropped notes, tuning by easy method of flattened fifths. No previous skills needed. 4 illustrations. 201pp. 5⅜ × 8½. 23267-0 Pa. $3.50

EARLY AMERICAN IRON-ON TRANSFER PATTERNS, edited by Rita Weiss. 75 designs, borders, alphabets, from traditional American sources. 48pp. 8¼ × 11. 23162-3 Pa. $1.95

CROCHETING EDGINGS, edited by Rita Weiss. Over 100 of the best designs for these lovely trims for a host of household items. Complete instructions, illustrations. 48pp. 8¼ × 11. 24031-2 Pa. $2.25

FINGER PLAYS FOR NURSERY AND KINDERGARTEN, Emilie Poulsson. 18 finger plays with music (voice and piano); entertaining, instructive. Counting, nature lore, etc. Victorian classic. 53 illustrations. 80pp. 6½ × 9¼. 22588-7 Pa. $1.95

BOSTON THEN AND NOW, Peter Vanderwarker. Here in 59 side-by-side views are photographic documentations of the city's past and present. 119 photographs. Full captions. 122pp. 8¼ × 11. 24312-5 Pa. $7.95

CROCHETING BEDSPREADS, edited by Rita Weiss. 22 patterns, originally published in three instruction books 1939-41. 39 photos, 8 charts. Instructions. 48pp. 8¼ × 11. 23610-2 Pa. $2.00

HAWTHORNE ON PAINTING, Charles W. Hawthorne. Collected from notes taken by students at famous Cape Cod School; hundreds of direct, personal *apercus*, ideas, suggestions. 91pp. 5⅜ × 8½. 20653-X Pa. $2.95

THERMODYNAMICS, Enrico Fermi. A classic of modern science. Clear, organized treatment of systems, first and second laws, entropy, thermodynamic potentials, etc. Calculus required. 160pp. 5⅜ × 8½. 60361-X Pa. $4.50

TEN BOOKS ON ARCHITECTURE, Vitruvius. The most important book ever written on architecture. Early Roman aesthetics, technology, classical orders, site selection, all other aspects. Morgan translation. 331pp. 5⅜ × 8½. 20645-9 Pa. $5.95

THE CORNELL BREAD BOOK, Clive M. McCay and Jeanette B. McCay. Famed high-protein recipe incorporated into breads, rolls, buns, coffee cakes, pizza, pie crusts, more. Nearly 50 illustrations. 48pp. 8¼ × 11. 23995-0 Pa. $2.00

THE CRAFTSMAN'S HANDBOOK, Cennino Cennini. 15th-century handbook, school of Giotto, explains applying gold, silver leaf; gesso; fresco painting, grinding pigments, etc. 142pp. 6⅛ × 9¼. 20054-X Pa. $3.50

FRANK LLOYD WRIGHT'S FALLINGWATER, Donald Hoffmann. Full story of Wright's masterwork at Bear Run, Pa. 100 photographs of site, construction, and details of completed structure. 112pp. 9¼ × 10. 23671-4 Pa. $7.95

OVAL STAINED GLASS PATTERN BOOK, C. Eaton. 60 new designs framed in shape of an oval. Greater complexity, challenge with sinuous cats, birds, mandalas framed in antique shape. 64pp. 8¼ × 11. 24519-5 Pa. $3.75

THE BOOK OF WOOD CARVING, Charles Marshall Sayers. Still finest book for beginning student. Fundamentals, technique; gives 34 designs, over 34 projects for panels, bookends, mirrors, etc. 33 photos. 118pp. 7¾ × 10⅝. 23654-4 Pa. $3.95

CARVING COUNTRY CHARACTERS, Bill Higginbotham. Expert advice for beginning, advanced carvers on materials, techniques for creating 18 projects—mirthful panorama of American characters. 105 illustrations. 80pp. 8⅜ × 11. 24135-1 Pa. $2.50

300 ART NOUVEAU DESIGNS AND MOTIFS IN FULL COLOR, C.B. Grafton. 44 full-page plates display swirling lines and muted colors typical of Art Nouveau. Borders, frames, panels, cartouches, dingbats, etc. 48pp. 9⅜ × 12¼. 24354-0 Pa. $6.95

SELF-WORKING CARD TRICKS, Karl Fulves. Editor of *Pallbearer* offers 72 tricks that work automatically through nature of card deck. No sleight of hand needed. Often spectacular. 42 illustrations. 113pp. 5⅜ × 8½. 23334-0 Pa. $3.50

CUT AND ASSEMBLE A WESTERN FRONTIER TOWN, Edmund V. Gillon, Jr. Ten authentic full-color buildings on heavy cardboard stock in H-O scale. Sheriff's Office and Jail, Saloon, Wells Fargo, Opera House, others. 48pp. 9¼ × 12¼. 23736-2 Pa. $4.95

CUT AND ASSEMBLE AN EARLY NEW ENGLAND VILLAGE, Edmund V. Gillon, Jr. Printed in full color on heavy cardboard stock. 12 authentic buildings in H-O scale: Adams home in Quincy, Mass., Oliver Wight house in Sturbridge, smithy, store, church, others. 48pp. 9¼ × 12¼. 23536-X Pa. $4.95

THE TALE OF TWO BAD MICE, Beatrix Potter. Tom Thumb and Hunca Munca squeeze out of their hole and go exploring. 27 full-color Potter illustrations. 59pp. 4¼ × 5½. (Available in U.S. only) 23065-1 Pa. $1.75

CARVING FIGURE CARICATURES IN THE OZARK STYLE, Harold L. Enlow. Instructions and illustrations for ten delightful projects, plus general carving instructions. 22 drawings and 47 photographs altogether. 39pp. 8⅜ × 11. 23151-8 Pa. $2.95

A TREASURY OF FLOWER DESIGNS FOR ARTISTS, EMBROIDERERS AND CRAFTSMEN, Susan Gaber. 100 garden favorites lushly rendered by artist for artists, craftsmen, needleworkers. Many form frames, borders. 80pp. 8¼ × 11. 24096-7 Pa. $3.50

CUT & ASSEMBLE A TOY THEATER/THE NUTCRACKER BALLET, Tom Tierney. Model of a complete, full-color production of Tchaikovsky's classic. 6 backdrops, dozens of characters, familiar dance sequences. 32pp. 9⅜ × 12¼. 24194-7 Pa. $4.50

ANIMALS: 1,419 COPYRIGHT-FREE ILLUSTRATIONS OF MAMMALS, BIRDS, FISH, INSECTS, ETC., edited by Jim Harter. Clear wood engravings present, in extremely lifelike poses, over 1,000 species of animals. 284pp. 9 × 12. 23766-4 Pa. $9.95

MORE HAND SHADOWS, Henry Bursill. For those at their 'finger ends," 16 more effects—Shakespeare, a hare, a squirrel, Mr. Punch, and twelve more—each explained by a full-page illustration. Considerable period charm. 30pp. 6½ × 9¼. 21384-6 Pa. $1.95

SURREAL STICKERS AND UNREAL STAMPS, William Rowe. 224 haunting, hilarious stamps on gummed, perforated stock, with images of elephants, geisha girls, George Washington, etc. 16pp. one side. 8¼ × 11. 24371-0 Pa. $3.50

GOURMET KITCHEN LABELS, Ed Sibbett, Jr. 112 full-color labels (4 copies each of 28 designs). Fruit, bread, other culinary motifs. Gummed and perforated. 16pp. 8¼ × 11. 24087-8 Pa. $2.95

PATTERNS AND INSTRUCTIONS FOR CARVING AUTHENTIC BIRDS, H.D. Green. Detailed instructions, 27 diagrams, 85 photographs for carving 15 species of birds so life-like, they'll seem ready to fly! 8¼ × 11. 24222-6 Pa. $2.75

FLATLAND, E.A. Abbott. Science-fiction classic explores life of 2-D being in 3-D world. 16 illustrations. 103pp. 5⅜ × 8. 20001-9 Pa. $2.00

DRIED FLOWERS, Sarah Whitlock and Martha Rankin. Concise, clear, practical guide to dehydration, glycerinizing, pressing plant material, and more. Covers use of silica gel. 12 drawings. 32pp. 5⅜ × 8½. 21802-3 Pa. $1.00

EASY-TO-MAKE CANDLES, Gary V. Guy. Learn how easy it is to make all kinds of decorative candles. Step-by-step instructions. 82 illustrations. 48pp. 8¼ × 11. 23881-4 Pa. $2.95

SUPER STICKERS FOR KIDS, Carolyn Bracken. 128 gummed and perforated full-color stickers: GIRL WANTED, KEEP OUT, BORED OF EDUCATION, X-RATED, COMBAT ZONE, many others. 16pp. 8¼ × 11. 24092-4 Pa. $2.50

CUT AND COLOR PAPER MASKS, Michael Grater. Clowns, animals, funny faces...simply color them in, cut them out, and put them together, and you have 9 paper masks to play with and enjoy. 32pp. 8¼ × 11. 23171-2 Pa. $2.50

A CHRISTMAS CAROL: THE ORIGINAL MANUSCRIPT, Charles Dickens. Clear facsimile of Dickens manuscript, on facing pages with final printed text. 8 illustrations by John Leech, 4 in color on covers. 144pp. 8⅜ × 11¼. 20980-6 Pa. $5.95

CARVING SHOREBIRDS, Harry V. Shourds & Anthony Hillman. 16 full-size patterns (all double-page spreads) for 19 North American shorebirds with step-by-step instructions. 72pp. 9¼ × 12¼. 24287-0 Pa. $4.95

THE GENTLE ART OF MATHEMATICS, Dan Pedoe. Mathematical games, probability, the question of infinity, topology, how the laws of algebra work, problems of irrational numbers, and more. 42 figures. 143pp. 5⅜ × 8½. (EBE) 22949-1 Pa. $3.50

READY-TO-USE DOLLHOUSE WALLPAPER, Katzenbach & Warren, Inc. Stripe, 2 floral stripes, 2 allover florals, polka dot; all in full color. 4 sheets (350 sq. in.) of each, enough for average room. 48pp. 8¼ × 11. 23495-9 Pa. $2.95

MINIATURE IRON-ON TRANSFER PATTERNS FOR DOLLHOUSES, DOLLS, AND SMALL PROJECTS, Rita Weiss and Frank Fontana. Over 100 miniature patterns: rugs, bedspreads, quilts, chair seats, etc. In standard dollhouse size. 48pp. 8¼ × 11. 23741-9 Pa. $1.95

THE DINOSAUR COLORING BOOK, Anthony Rao. 45 renderings of dinosaurs, fossil birds, turtles, other creatures of Mesozoic Era. Scientifically accurate. Captions. 48pp. 8¼ × 11. 24022-3 Pa. $2.50

CHANCERY CURSIVE STROKE BY STROKE, Arthur Baker. Instructions and illustrations for each stroke of each letter (upper and lower case) and numerals. 54 full-page plates. 64pp. 8¼ × 11. 24278-1 Pa. $2.50

THE ENJOYMENT AND USE OF COLOR, Walter Sargent. Color relationships, values, intensities; complementary colors, illumination, similar topics. Color in nature and art. 7 color plates, 29 illustrations. 274pp. 5⅜ × 8½. 20944-X Pa. $4.95

SCULPTURE PRINCIPLES AND PRACTICE, Louis Slobodkin. Step-by-step approach to clay, plaster, metals, stone; classical and modern. 253 drawings, photos. 255pp. 8⅛ × 11. 22960-2 Pa. $7.50

VICTORIAN FASHION PAPER DOLLS FROM HARPER'S BAZAR, 1867-1898, Theodore Menten. Four female dolls with 28 elegant high fashion costumes, printed in full color. 32pp. 9¼ × 12¼. (USCO) 23453-3 Pa. $3.95

FLOPSY, MOPSY AND COTTONTAIL: A Little Book of Paper Dolls in Full Color, Susan LaBelle. Three dolls and 21 costumes (7 for each doll) show Peter Rabbit's siblings dressed for holidays, gardening, hiking, etc. Charming borders, captions. 48pp. 4¼ × 5½. 24376-1 Pa. $2.50

NATIONAL LEAGUE BASEBALL CARD CLASSICS, Bert Randolph Sugar. 83 big-leaguers from 1909-69 on facsimile cards. Hubbell, Dean, Spahn, Brock plus advertising, info, no duplications. Perforated, detachable. 16pp. 8¼ × 11. 24308-7 Pa. $2.95

THE LOGICAL APPROACH TO CHESS, Dr. Max Euwe, et al. First-rate text of comprehensive strategy, tactics, theory for the amateur. No gambits to memorize, just a clear, logical approach. 224pp. 5⅜ × 8½. 24353-2 Pa. $4.50

MAGICK IN THEORY AND PRACTICE, Aleister Crowley. The summation of the thought and practice of the century's most famous necromancer, long hard to find. Crowley's best book. 436pp. 5⅜ × 8½. (Available in U.S. only) 23295-6 Pa. $6.50

THE HAUNTED HOTEL, Wilkie Collins. Collins' last great tale; doom and destiny in a Venetian palace. Praised by T.S. Eliot. 127pp. 5⅜ × 8½. 24333-8 Pa. $3.00

ART DECO DISPLAY ALPHABETS, Dan X. Solo. Wide variety of bold yet elegant lettering in handsome Art Deco styles. 100 complete fonts, with numerals, punctuation, more. 104pp. 8⅛ × 11. 24372-9 Pa. $4.50

CALLIGRAPHIC ALPHABETS, Arthur Baker. Nearly 150 complete alphabets by outstanding contemporary. Stimulating ideas; useful source for unique effects. 154 plates. 157pp. 8⅜ × 11¼. 21045-6 Pa. $5.95

ARTHUR BAKER'S HISTORIC CALLIGRAPHIC ALPHABETS, Arthur Baker. From monumental capitals of first-century Rome to humanistic cursive of 16th century, 33 alphabets in fresh interpretations. 88 plates. 96pp. 9 × 12. 24054-1 Pa. $4.50

LETTIE LANE PAPER DOLLS, Sheila Young. Genteel turn-of-the-century family very popular then and now. 24 paper dolls. 16 plates in full color. 32pp. 9¼ × 12¼. 24089-4 Pa. $3.50

KEYBOARD WORKS FOR SOLO INSTRUMENTS, G.F. Handel. 35 neglected works from Handel's vast oeuvre, originally jotted down as improvisations. Includes Eight Great Suites, others. New sequence. 174pp. 9⅜ × 12¼.
24338-9 Pa. $7.50

AMERICAN LEAGUE BASEBALL CARD CLASSICS, Bert Randolph Sugar. 82 stars from 1900s to 60s on facsimile cards. Ruth, Cobb, Mantle, Williams, plus advertising, info, no duplications. Perforated, detachable. 16pp. 8¼ × 11.
24286-2 Pa. $2.95

A TREASURY OF CHARTED DESIGNS FOR NEEDLEWORKERS, Georgia Gorham and Jeanne Warth. 141 charted designs: owl, cat with yarn, tulips, piano, spinning wheel, covered bridge, Victorian house and many others. 48pp. 8¼ × 11.
23558-0 Pa. $1.95

DANISH FLORAL CHARTED DESIGNS, Gerda Bengtsson. Exquisite collection of over 40 different florals: anemone, Iceland poppy, wild fruit, pansies, many others. 45 illustrations. 48pp. 8¼ × 11. 23957-8 Pa. $1.95

OLD PHILADELPHIA IN EARLY PHOTOGRAPHS 1839-1914, Robert F. Looney. 215 photographs: panoramas, street scenes, landmarks, President-elect Lincoln's visit, 1876 Centennial Exposition, much more. 230pp. 8⅜ × 11¾.
23345-6 Pa. $9.95

PRELUDE TO MATHEMATICS, W.W. Sawyer. Noted mathematician's lively, stimulating account of non-Euclidean geometry, matrices, determinants, group theory, other topics. Emphasis on novel, striking aspects. 224pp. 5⅜ × 8½.
24401-6 Pa. $4.50

ADVENTURES WITH A MICROSCOPE, Richard Headstrom. 59 adventures with clothing fibers, protozoa, ferns and lichens, roots and leaves, much more. 142 illustrations. 232pp. 5⅜ × 8½. 23471-1 Pa. $3.95

IDENTIFYING ANIMAL TRACKS: MAMMALS, BIRDS, AND OTHER ANIMALS OF THE EASTERN UNITED STATES, Richard Headstrom. For hunters, naturalists, scouts, nature-lovers. Diagrams of tracks, tips on identification. 128pp. 5⅜ × 8. 24442-3 Pa. $3.50

VICTORIAN FASHIONS AND COSTUMES FROM HARPER'S BAZAR, 1867-1898, edited by Stella Blum. Day costumes, evening wear, sports clothes, shoes, hats, other accessories in over 1,000 detailed engravings. 320pp. 9⅜ × 12¼.
22990-4 Pa. $10.95

EVERYDAY FASHIONS OF THE TWENTIES AS PICTURED IN SEARS AND OTHER CATALOGS, edited by Stella Blum. Actual dress of the Roaring Twenties, with text by Stella Blum. Over 750 illustrations, captions. 156pp. 9 × 12.
24134-3 Pa. $8.50

HALL OF FAME BASEBALL CARDS, edited by Bert Randolph Sugar. Cy Young, Ted Williams, Lou Gehrig, and many other Hall of Fame greats on 92 full-color, detachable reprints of early baseball cards. No duplication of cards with *Classic Baseball Cards*. 16pp. 8¼ × 11. 23624-2 Pa. $3.50

THE ART OF HAND LETTERING, Helm Wotzkow. Course in hand lettering, Roman, Gothic, Italic, Block, Script. Tools, proportions, optical aspects, individual variation. Very quality conscious. Hundreds of specimens. 320pp. 5⅜ × 8½.
21797-3 Pa. $4.95

HOW THE OTHER HALF LIVES, Jacob A. Riis. Journalistic record of filth, degradation, upward drive in New York immigrant slums, shops, around 1900. New edition includes 100 original Riis photos, monuments of early photography. 233pp. 10 × 7⅞. 22012-5 Pa. $7.95

CHINA AND ITS PEOPLE IN EARLY PHOTOGRAPHS, John Thomson. In 200 black-and-white photographs of exceptional quality photographic pioneer Thomson captures the mountains, dwellings, monuments and people of 19th-century China. 272pp. 9⅜ × 12¼. 24393-1 Pa. $13.95

GODEY COSTUME PLATES IN COLOR FOR DECOUPAGE AND FRAMING, edited by Eleanor Hasbrouk Rawlings. 24 full-color engravings depicting 19th-century Parisian haute couture. Printed on one side only. 56pp. 8¼ × 11. 23879-2 Pa. $3.95

ART NOUVEAU STAINED GLASS PATTERN BOOK, Ed Sibbett, Jr. 104 projects using well-known themes of Art Nouveau: swirling forms, florals, peacocks, and sensuous women. 60pp. 8¼ × 11. 23577-7 Pa. $3.50

QUICK AND EASY PATCHWORK ON THE SEWING MACHINE: Susan Aylsworth Murwin and Suzzy Payne. Instructions, diagrams show exactly how to machine sew 12 quilts. 48pp. of templates. 50 figures. 80pp. 8¼ × 11. 23770-2 Pa. $3.50

THE STANDARD BOOK OF QUILT MAKING AND COLLECTING, Marguerite Ickis. Full information, full-sized patterns for making 46 traditional quilts, also 150 other patterns. 483 illustrations. 273pp. 6⅞ × 9⅝. 20582-7 Pa. $5.95

LETTERING AND ALPHABETS, J. Albert Cavanagh. 85 complete alphabets lettered in various styles; instructions for spacing, roughs, brushwork. 121pp. 8¾ × 8. 20053-1 Pa. $3.95

LETTER FORMS: 110 COMPLETE ALPHABETS, Frederick Lambert. 110 sets of capital letters; 16 lower case alphabets; 70 sets of numbers and other symbols. 110pp. 8⅛ × 11. 22872-X Pa. $4.50

ORCHIDS AS HOUSE PLANTS, Rebecca Tyson Northen. Grow cattleyas and many other kinds of orchids—in a window, in a case, or under artificial light. 63 illustrations. 148pp. 5⅜ × 8½. 23261-1 Pa. $2.95

THE MUSHROOM HANDBOOK, Louis C.C. Krieger. Still the best popular handbook. Full descriptions of 259 species, extremely thorough text, poisons, folklore, etc. 32 color plates; 126 other illustrations. 560pp. 5⅜ × 8½. 21861-9 Pa. $8.50

THE DORÉ BIBLE ILLUSTRATIONS, Gustave Doré. All wonderful, detailed plates: Adam and Eve, Flood, Babylon, life of Jesus, etc. Brief King James text with each plate. 241 plates. 241pp. 9 × 12. 23004-X Pa. $8.95

THE BOOK OF KELLS: Selected Plates in Full Color, edited by Blanche Cirker. 32 full-page plates from greatest manuscript-icon of early Middle Ages. Fantastic, mysterious. Publisher's Note. Captions. 32pp. 9¾ × 12¼. 24345-1 Pa. $4.50

THE PERFECT WAGNERITE, George Bernard Shaw. Brilliant criticism of the Ring Cycle, with provocative interpretation of politics, economic theories behind the Ring. 136pp. 5⅜ × 8½. (EUK) 21707-8 Pa. $3.00

THE RIME OF THE ANCIENT MARINER, Gustave Doré, S.T. Coleridge. Doré's finest work, 34 plates capture moods, subtleties of poem. Full text. 77pp. 9¼ × 12. 22305-1 Pa. $4.95

SONGS OF INNOCENCE, William Blake. The first and most popular of Blake's famous "Illuminated Books," in a facsimile edition reproducing all 31 brightly colored plates. Additional printed text of each poem. 64pp. 5¼ × 7. 22764-2 Pa. $3.50

AN INTRODUCTION TO INFORMATION THEORY, J.R. Pierce. Second (1980) edition of most impressive non-technical account available. Encoding, entropy, noisy channel, related areas, etc. 320pp. 5⅜ × 8½. 24061-4 Pa. $4.95

THE DIVINE PROPORTION: A STUDY IN MATHEMATICAL BEAUTY, H.E. Huntley. "Divine proportion" or "golden ratio" in poetry, Pascal's triangle, philosophy, psychology, music, mathematical figures, etc. Excellent bridge between science and art. 58 figures. 185pp. 5⅜ × 8½. 22254-3 Pa. $3.95

THE DOVER NEW YORK WALKING GUIDE: From the Battery to Wall Street, Mary J. Shapiro. Superb inexpensive guide to historic buildings and locales in lower Manhattan: Trinity Church, Bowling Green, more. Complete Text; maps. 36 illustrations. 48pp. 3⅞ × 9¼. 24225-0 Pa. $2.50

NEW YORK THEN AND NOW, Edward B. Watson, Edmund V. Gillon, Jr. 83 important Manhattan sites: on facing pages early photographs (1875-1925) and 1976 photos by Gillon. 172 illustrations. 171pp. 9¼ × 10. 23361-8 Pa. $9.95

HISTORIC COSTUME IN PICTURES, Braun & Schneider. Over 1450 costumed figures from dawn of civilization to end of 19th century. English captions. 125 plates. 256pp. 8⅜ × 11¼. 23150-X Pa. $7.50

VICTORIAN AND EDWARDIAN FASHION: A Photographic Survey, Alison Gernsheim. First fashion history completely illustrated by contemporary photographs. Full text plus 235 photos, 1840-1914, in which many celebrities appear. 240pp. 6½ × 9¼. 24205-6 Pa. $6.00

CHARTED CHRISTMAS DESIGNS FOR COUNTED CROSS-STITCH AND OTHER NEEDLECRAFTS, Lindberg Press. Charted designs for 45 beautiful needlecraft projects with many yuletide and wintertime motifs. 48pp. 8¼ × 11. (EDNS) 24356-7 Pa. $2.50

101 FOLK DESIGNS FOR COUNTED CROSS-STITCH AND OTHER NEEDLECRAFTS, Carter Houck. 101 authentic charted folk designs in a wide array of lovely representations with many suggestions for effective use. 48pp. 8¼ × 11. 24369-9 Pa. $2.25

FIVE ACRES AND INDEPENDENCE, Maurice G. Kains. Great back-to-the-land classic explains basics of self-sufficient farming. The one book to get. 95 illustrations. 397pp. 5⅜ × 8½. 20974-1 Pa. $5.95

A MODERN HERBAL, Margaret Grieve. Much the fullest, most exact, most useful compilation of herbal material. Gigantic alphabetical encyclopedia, from aconite to zedoary, gives botanical information, medical properties, folklore, economic uses, and much else. Indispensable to serious reader. 161 illustrations. 888pp. 6½ × 9¼. (Available in U.S. only) 22798-7, 22799-5 Pa., Two-vol. set $16.45

DECORATIVE NAPKIN FOLDING FOR BEGINNERS, Lillian Oppenheimer and Natalie Epstein. 22 different napkin folds in the shape of a heart, clown's hat, love knot, etc. 63 drawings. 48pp. 8¼ × 11. 23797-4 Pa. $1.95

DECORATIVE LABELS FOR HOME CANNING, PRESERVING, AND OTHER HOUSEHOLD AND GIFT USES, Theodore Menten. 128 gummed, perforated labels, beautifully printed in 2 colors. 12 versions. Adhere to metal, glass, wood, ceramics. 24pp. 8¼ × 11. 23219-0 Pa. $3.50

EARLY AMERICAN STENCILS ON WALLS AND FURNITURE, Janet Waring. Thorough coverage of 19th-century folk art: techniques, artifacts, surviving specimens. 166 illustrations, 7 in color. 147pp. of text. 7⅞ × 10¾. 21906-2 Pa. $9.95

AMERICAN ANTIQUE WEATHERVANES, A.B. & W.T. Westervelt. Extensively illustrated 1883 catalog exhibiting over 550 copper weathervanes and finials. Excellent primary source by one of the principal manufacturers. 104pp. 6⅛ × 9¼. 24396-6 Pa. $3.95

ART STUDENTS' ANATOMY, Edmond J. Farris. Long favorite in art schools. Basic elements, common positions, actions. Full text, 158 illustrations. 159pp. 5⅜ × 8½. 20744-7 Pa. $3.95

BRIDGMAN'S LIFE DRAWING, George B. Bridgman. More than 500 drawings and text teach you to abstract the body into its major masses. Also specific areas of anatomy. 192pp. 6½ × 9¼. (EA) 22710-3 Pa. $4.50

COMPLETE PRELUDES AND ETUDES FOR SOLO PIANO, Frederic Chopin. All 26 Preludes, all 27 Etudes by greatest composer of piano music. Authoritative Paderewski edition. 224pp. 9 × 12. (Available in U.S. only) 24052-5 Pa. $7.50

PIANO MUSIC 1888-1905, Claude Debussy. Deux Arabesques, Suite Bergamesque, Masques, 1st series of Images, etc. 9 others, in corrected editions. 175pp. 9⅜ × 12¼. 22771-5 Pa. $5.95

TEDDY BEAR IRON-ON TRANSFER PATTERNS, Ted Menten. 80 iron-on transfer patterns of male and female Teddys in a wide variety of activities, poses, sizes. 48pp. 8¼ × 11. 24596-9 Pa. $2.25

A PICTURE HISTORY OF THE BROOKLYN BRIDGE, M.J. Shapiro. Profusely illustrated account of greatest engineering achievement of 19th century. 167 rare photos & engravings recall construction, human drama. Extensive, detailed text. 122pp. 8¼ × 11. 24403-2 Pa. $7.95

NEW YORK IN THE THIRTIES, Berenice Abbott. Noted photographer's fascinating study shows new buildings that have become famous and old sights that have disappeared forever. 97 photographs. 97pp. 11⅝ × 10. 22967-X Pa. $7.50

MATHEMATICAL TABLES AND FORMULAS, Robert D. Carmichael and Edwin R. Smith. Logarithms, sines, tangents, trig functions, powers, roots, reciprocals, exponential and hyperbolic functions, formulas and theorems. 269pp. 5⅜ × 8½. 60111-0 Pa. $4.95

HANDBOOK OF MATHEMATICAL FUNCTIONS WITH FORMULAS, GRAPHS, AND MATHEMATICAL TABLES, edited by Milton Abramowitz and Irene A. Stegun. Vast compendium: 29 sets of tables, some to as high as 20 places. 1,046pp. 8 × 10½. 61272-4 Pa. $19.95

CATALOG OF DOVER BOOKS

SOURCE BOOK OF MEDICAL HISTORY, edited by Logan Clendening, M.D. Original accounts ranging from Ancient Egypt and Greece to discovery of X-rays: Galen, Pasteur, Lavoisier, Harvey, Parkinson, others. 685pp. 5⅜ × 8½.
20621-1 Pa. $10.95

THE ROSE AND THE KEY, J.S. Lefanu. Superb mystery novel from Irish master. Dark doings among an ancient and aristocratic English family. Well-drawn characters; capital suspense. Introduction by N. Donaldson. 448pp. 5⅜ × 8½.
24377-X Pa. $6.95

SOUTH WIND, Norman Douglas. Witty, elegant novel of ideas set on languorous Meditterranean island of Nepenthe. Elegant prose, glittering epigrams, mordant satire. 1917 masterpiece. 416pp. 5⅜ × 8½. (Available in U.S. only)
24361-3 Pa. $5.95

RUSSELL'S CIVIL WAR PHOTOGRAPHS, Capt. A.J. Russell. 116 rare Civil War Photos: Bull Run, Virginia campaigns, bridges, railroads, Richmond, Lincoln's funeral car. Many never seen before. Captions. 128pp. 9⅜ × 12¼.
24283-8 Pa. $7.95

PHOTOGRAPHS BY MAN RAY: 105 Works, 1920-1934. Nudes, still lifes, landscapes, women's faces, celebrity portraits (Dali, Matisse, Picasso, others), rayographs. Reprinted from rare gravure edition. 128pp. 9⅜ × 12¼. (Available in U.S. only)
23842-3 Pa. $7.95

STAR NAMES: THEIR LORE AND MEANING, Richard H. Allen. Star names, the zodiac, constellations: folklore and literature associated with heavens. The basic book of its field, fascinating reading. 563pp. 5⅜ × 8½. 21079-0 Pa. $7.95

BURNHAM'S CELESTIAL HANDBOOK, Robert Burnham, Jr. Thorough guide to the stars beyond our solar system. Exhaustive treatment. Alphabetical by constellation: Andromeda to Cetus in Vol. 1; Chamaeleon to Orion in Vol. 2; and Pavo to Vulpecula in Vol. 3. Hundreds of illustrations. Index in Vol. 3. 2000pp. 6⅛ × 9¼. 23567-X, 23568-8, 23673-0 Pa. Three-vol. set $36.85

THE ART NOUVEAU STYLE BOOK OF ALPHONSE MUCHA, Alphonse Mucha. All 72 plates from *Documents Decoratifs* in original color. Stunning, essential work of Art Nouveau. 80pp. 9⅜ × 12¼. 24044-4 Pa. $7.95

DESIGNS BY ERTE; FASHION DRAWINGS AND ILLUSTRATIONS FROM "HARPER'S BAZAR," Erte. 310 fabulous line drawings and 14 *Harper's Bazar* covers, 8 in full color. Erte's exotic temptresses with tassels, fur muffs, long trains, coifs, more. 129pp. 9⅜ × 12¼. 23397-9 Pa. $6.95

HISTORY OF STRENGTH OF MATERIALS, Stephen P. Timoshenko. Excellent historical survey of the strength of materials with many references to the theories of elasticity and structure. 245 figures. 452pp. 5⅜ × 8½. 61187-6 Pa. $8.95

Prices subject to change without notice.

Available at your book dealer or write for free catalog to Dept. GI, Dover Publications, Inc., 31 East 2nd St. Mineola, N.Y. 11501. Dover publishes more than 175 books each year on science, elementary and advanced mathematics, biology, music, art, literary history, social sciences and other areas.